ENVIRONMENTAL POLLUTION

About the Author

R.D. Gupta, M.Sc. and Ph. D. (Agriculture), is a well-known soil scientist and environmentalist, having specialisation in soil microbiology and pedology, and in wildlife of Pir Panjal Himalayan region of Jammu and Kashmir. He did his postgraduate diploma in ecology and environment from Indian Institute of Ecology and Environment, New Delhi.

Starting his professional career as Assistant Extension Specialist (Soil Science) in Punjab Agricultural University, Ludhiana in 1968, Prof. Gupta served as Associate Professor, Deputy Director, Extension Education (Training) in Chaudhary Sarwan Kumar Himachal Pradesh Krishi Vishwavidyalaya and Sher-e-Kashmir University of Agricultural Sciences and Technology, before he was elevated as Chief Scientist, KVK and Head SKUAST, Regional Agricultural Research Station, R.S. Pura, Jammu. Later, he served as Founding Associate Dean, Faculty of Agriculture, SKUAST, Jammu; and Head, Divisions of Agricultural Chemistry, and Soil Science, and Agro-forestry. He has to his credit over 100 research papers and book, entitled, *Problems and Management of Soil and Forest Resources of Northwest Himalayas*.

ENVIRONMENTAL POLLUTION
HAZARDS AND CONTROL

R.D. GUPTA

Associate Dean (Retd.), Faculty of Agriculture, SKUAST – Jammu, J&K State

CONCEPT PUBLISHING COMPANY, NEW DELHI-110059

Cataloging in Publication Data—DK
Courtesy: D.K. Agencies (P) Ltd. <docinfo@dkagencies.com>

Gupta, R. D. (Rameshwar Dass), 1941-
Environmental pollution : hazards and control / R.D. Gupta.
p. cm.
Includes bibliographical references (p.
Includes index.
ISBN 8180692604

1. Pollution—India. I. Title.

DDC 363.730 954 22

ISBN 81-8069-260-4

First Published 2006

Published and Printed by
Ashok Kumar Mittal
Concept Publishing Company
A/15-16, Commercial Block, Mohan Garden,
New Delhi-110059 (India)

Phones: 25351460, 25351794
Fax: 091-(11)-25357103
Email : publishing@conceptpub.com

FOREWORD

The word "environment" refers to surroundings which vary from place to place and continent to continent depending upon physiography, topography, climate and natural resources. Since the beginning of the culture, natural resources like land/soil, air, water, forests, petroleum, stones and minerals have been exploited in different ways. However, at any moment when these natural resources are over exploited the environment gets polluted, damaged or degraded.

Because of the burgeoning population of human beings and livestock, the forest and grassland resources have been overexploited. This has resulted in multidimensional environmental crisis like denudation, soil erosion, landslides and flash floods. These in turn have created dust pollution, soil pollution and metal pollution. Tremendous increase in traffic and industrial sector have produced vehicular pollution and air pollution. Infact, vehicular emission is contributing to air pollution substantially *vis-à-vis* sound or noise pollution. Indiscriminate use of pesticides and fertilizers have posed pesticidal pollution, soil pollution and food pollution. Various industrial effluents in conjunction with organic wastes and washings of pesticide and fertilizer residues during rains and/or irrigation water, are mainly amenable to cause water pollution and metal pollution. Recently, all kinds of bottled water whether national (like Bisleri) or multinational (Kinley) and soft drinks like Coca Cola, Pepsi etc. have been found contaminated with pesticide residues, which is of great concern for health hazards.

All the aforesaid kinds of environment pollutions are playing havoc not only to human beings and animals but also to vegetative kingdom. For instance, the impact of air pollution on health is considerable. Premature deaths due to respiratory and cardio-vascular diseases and illness due to chronic respiratory diseases like asthma and bronchitis have increased. Lead emitted in the atmosphere from auto exhaust causes lead poisoning of human body leading to stomach ailments, indigestion, constipation, lack of appetite and sleeplessness. Dust from blast mining and grinding, stone crushers and cement factories, leaves trace elements in the atmosphere like cobalt, nickel, chromium, selenium, fluorides, calcium and magnesium. These elements have proved harmful to human beings besides affecting the flora and fauna. The highly toxic effluents and residues of pesticides and fertilizers, are being regularly added in rivers, lakes and streams making their water unfit for drinking and bathing purposes. Water polluted with heavy metals like mercury, lead, cadmium, chromium is more dangerous than the municipal wastes. Excess of fluoride causes fluorosis, a disease that cripples human beings. Excess of nitrate creates blue baby diseases in infants. Noise pollution is another health hazard which is causing tension, headache, deafness and many other diseases.

Recent reports indicate that about 70 per cent of the cereals, milk and milk products are contaminated with pesticide residues. Toxic chemicals used for preserving various foods have not only caused food pollution but air pollution also. Indiscriminate use of fertilizers has adversely affected the soil ecological balance. The population of earthworms is vanishing. Similar is the case with soil nitrifiers and nitrogen fixers. Milk for the infants is too contaminated/adulterated which will cause more harm to them. Breast feeding has been almost abandoned. All these factors have aggravated the life hazards especially of young ones. After all science is meant to make life more healthy and pleasant. How, it can be done is a challenge to the scientists. They have to educate the ignorant masses of this large country. It is their responsibility to create awareness in simple language and infact put them in religious and social temper so that the rural poor or slum dwellers who are being affected the most also understand and participate in dealing with pollutions. All the remedies and awareness programmes given in the book must be converted into mantras/slogans like *Sabaz patta jalana pap hai* (to burn green leaf is a sin). In this direction, this book is a great contribution.

In view of the hazards created by the environmental pollution as stated above, it becomes imperative to bring them out in more details in the form of a book alongwith their control measures for understanding of the urban and rural masses. So I fully appreciate the idea of publishing the book entitled *Environmental Pollution : Hazards and Control* by Prof. R.D. Gupta, which is need of the hour. The contents of the book have been divided into 11 chapters. They cover introduction, air, water and noise pollutions as well as vehicular, metal, dust, soil, pesticidal, radioactive and food pollutions. A number of topical issues which call for immediate heed to tackle, are detailed in this book. They are air pollution in India, fresh water crisis in India, noise absorbant plant species, noise pollution - legal prospectives, legal implications amenable for vehicular pollution, impact of acid rains on agricultural lands, water resources; forest resources; animals and wildlife and human health, botanical pesticides and microbial control for checking insects, pests, effects of radioactive pollutants, adulterants and natural organic toxicants. Moreover, control measures based on ecological principles given separately for each chapter, and priorities for the future, are the other crucial topics.

I think the book fulfills in our current knowledge of the "Environmental Pollution", whose impact on the human and animal population as well as on inanimate objects of relevance to Indian conditions which have not received the attention it deserves. I feel this book will add to the growing interest of Indian scientists from data of different chapters.

It is hoped that this book augurs well for the future of the emerging new field of environmental pollution in the environment ecology. Infact, I would advise all scientists to invent ideas, schemes and programmes, from the control measures/suggestions given in this book, which can be adopted by the rural people to augment their income. These programmes may include agroforestry, eco-development and watershed management to increase production in agriculture and allied subjects *vis-a-vis* to conserve soil and water.

— **Major General Goverdhan Singh Jamwal (Retd.), AVSM**

Ex-Member J&K Legislative Council, presently President, J&K Paryavaran (Environment) Sanstha, Gadhadhar Mandir, Jammu.

PREFACE

In the relentless march towards development by means of environmentally incompatible industrialisation and urbanisation, natural resources such as mines, soils, forests, grasslands and rangelands are being plundered without sufficient thought for the long term consequences of such actions. Lopsided orientation of Science and Technology, population explosion as well as our sort sighted and selfish interests have caused a number of environmental problems.

The process of industrialisation has resulted in rapid degradation of human environment. Today India has serious environmental problems. The main areas of concern fall into two categories (a) pollution issues which are largely urban and industrial based and (b) resource management issues which consist of water, soils, forests, pastures, rangelands and coast zone management *vis-a-vis* biodiversity.

On the pollution side urban air, water and solid waste problems are leading to significant health impacts. The people working in the fields, factories and mines are exposed to dust, fumes of vapours of different chemicals which cause hitherto unknown occupational health hazards. They are also subjected to heavy noise pollution and heat stress. Indiscriminate use of pesticides and fertilizers for increasing agriculture production have caused pesticidal and food pollution. Heavy metal pollution of soil and water is enormously increasing and has now become a major environmental menace due to intensive mining, large production of sewerage and night soil, burning of fossil fuels like coal, petroleum.

On the resource side, the crop yields are declining because of soil pollution as well as soil erosion and salinity/sodicity. Forests have depleted and degraded for agricultural, livestock and fuel wood purposes. Rangelands are increasingly deteriorating. Due to heavy grazing pasture lands have become degraded. Because of degraded nature of forests, range and pasture lands, country's rich floral and faunal species are in danger of extinction.

No doubt, lot of information is available with regard to environmental pollution but it is lopsided and fragmentary. This information either pertains to soil pollution or pesticidal pollution only. In some cases, it deals with merely air or water pollution. But in light of more deterioration of the environment as elucidated above, the need of the hour is to throw light on various kinds of pollution in single volume as an integrated approach. Moreover, no book is available so far on environmental pollution for Indian students with data of our own country. Hence the present book entitled *Environmental Pollution: Hazards and Control*, covering a number of chapters concerning with various kinds of pollution, is first of its type.

The present volume of the book possesses eleven chapters. The first chapter deals with introduction which refers to about environment, change of ecological balance and pollution, reasons for arising environmental pollution, awareness of pollution and creation of Environment Ministry in India and impact of environmental pollution.

The second, third and fourth chapters focus on air, water and noise pollutions respectively. The objective here is to provide the students with an understanding of the factors or sources amenable for causing air, water and noise pollutions, their health hazards and control measures *vis-a-vis* priorities for the future in these lines.

Vehicular pollution in terms of its definition, status in India, causes, effects on human and animal life and management has been detailed in chapter five alongwith priorities for future, while in chapter six sources of metal pollution, its health hazards and management, are described. Chapter seven covers the causes of health hazards and management of dust pollution.

Various causes responsible for producing soil pollution, pesticidal pollution and radioactive pollution, are skilfully presented in chapters eight, nine and ten respectively. Effects of these pollution on human health, plant and animal life as well as their management too have been given. Chapter eleven covers food pollution, comprising food contamination and adulteration, causes and hazards.

It is hoped that the present volume, will prove useful for students, research scholars, teachers, environmentalists and extensionists by way of contributing a better understanding of the environmental problems.

R.D. Gupta,

39, Dhakki Sarajan,
Jammu 180001
Jammu & Kashmir
State, India

ACKNOWLEDGEMENTS

I express my sincere thanks to Major General Goverdhan Singh Jamwal (Retd.), AVSM, Ex-Member J&K Legislative Council, presently President, Jammu & Kashmir Paryavaran (Environment) Sanstha, Jammu for issuance of foreword after going through the manuscript critically and imparting healthy and valuable suggestions.

I also wish to thank Sh. Sohan Singh, IFS, former Chief Conservator of Forests (Retd.) Jammu and Kashmir Government for his unfailing and untiring help in going through the manuscript critically and imparting necessary corrections and suggestions. The heartfelt thanks are also due to all the Scientists and Philosophers whose work has been referred to in this book.

Sincere thanks are due to Sh. J.K. Verma, Ex-Executive Engineer (Retd.) Jammu and Kashmir Government for his cooperation in compiling the data. I am also thankful to Dr. R.M. Sharma, Assistant Professor, Pomology, SKUAST, Jammu for his sound criticism and help from time to time in compilation of the book. The help rendered by Dr. S.K. Gupta, Associate Professor, Agroforestry and Sh. A.K. Sharma, Assistant Professor Agronomy, SKUAST, Jammu in providing data relevant to book is acknowledged with thanks.

I am indebted to my brother Shri Satya Paul Mahajan and nephew Mr. Shambu Nath for inspiring the documentation of this book. Although both have left for heavenly abode in an illfate accident, yet the encouragement during their life period proved a milestone in bringing this book out.

R.D. Gupta

CONTENTS

1
INTRODUCTION

Environment, Ecological Balance and Environmental Pollution

The earth – a unique gift of God – is the only known planet which is inhabited by life. Life on earth was originated and flourished only because the environment of this planet provides necessary conditions for its sustenance. The word environment, the basis for human survival, refers to 'a surrounding'. It is an established fact that there exists a vital link between the environment and life. Every living organism thrives best in the amidst of biotic (plants and animals) and non-biotic (land, air, water) components, events, influences, the aggregate of which constitutes the environment. The environment, thus embraces sum total of all conditions and influences which affect the development and life of all organisms viz. microbes, plants and animals including human beings.

A natural ecological setting or stable ecological balance is required for healthy existence and preservation of the essential ingredients of life. But this balance is being upset by misuse, abuse and uncontrolled use of the environmental resources, thereby, endangering the very existence of, *inter alia*, human race. A man himself is amenable for drastically changing ecological balance through his intelligence, science and technology. Armed with the power of science and technology, he has tried to control his own evolutionary process. This has eventually resulted in the degradation of the environment which is his life support system. Hence, the alterations or changes which are responsible for degradation of the environment, have been termed as 'environmental pollution'. It can be defined as 'an undesirable change in the physical, chemical or biological characteristics of land, air and water that may or will harmfully affect human life or those of desirable species, industrial processes, living conditions and cultural assets'. In simple words, pollution can be stated 'any change in the environment leading to discomforts to human, animal and plant life'.

Reasons for arising Environmental Pollution

Basically the environmental pollution that threatens the future of mankind, is the gift of industrial revolution *vis-a-vis* over population of human beings and livestock, agricultural operations, domestic activities and illiteracy. In other words, we can say that environmental pollution is the byproduct of modern civilization which is impacting environment in various ways all over the world.

Growth of industries is generally considered as an indication of progress and prosperity but also brings deterioration to environmental health. The air, which is most utilized natural resource, is heavily polluted in many areas due to industrial smoke as well as smoke from heating units in households. The alarming dimensions of rising air pollution in India are evident from the fact that country's capital, Delhi has emerged as the fourth most polluted city in the world. Due to this reason, Delhi is bound to show skyrocketing incidence of the lung cancer. Similarly, water pollution is a worldwide problem and is more acute in the industrialised countries where millions of tonnes of pollutants are discharged in the rivers and streams and directly into the sea. Presently, 70 per cent rivers and streams of India contain polluted water which is about 10 times more than 70 years ago. High intensity sound or noise generated by industrial establishments, defence materials and practices, air travel, automobiles and domestic appliances, affecting human wellbeing and communications, is also a main problem especially in big cities like Delhi, Mumbai, Chennai, Kolkata and many others.

Prior to industrialisation, the agrarian cultures created environmental pollution in the form of soil erosion and landslides/landslips through deforestation and overgrazing of the pastures. An increase in population beyond carrying capacity of the environment, has overexploited of the natural resources, which in turn resulted in air, water and soil pollution as well as scarcity of food and poor living standards. Land resources fundamental components of the life supporting system are highly degraded by soil erosion, salinization, alkalization, acidification and water logging. These have aggravated by human activities and are serious threat to land as resource. Contamination of soil caused by human and livestock excreta is a serious health hazard.

Awareness of Pollution and Creation of Environment Ministry in India

The developed countries became aware of the pollution problems much earlier and good deal of work has been done to contain environmental pollution. In our country, the attention towards environmental pollution has been drawn recently. Even in the early decade of twentieth century, the Indians were not much concerned and sensitive about the environmental problems. In India, Ministry of Environment and Forests came into existence in 1980 and, thereafter, State Governments also created Directorate of Environment and Ecology. So, Central and State Governments are deeply committed to safeguard the environment. As a matter of fact, environment awareness is not new in India but protection of environment is our constitutional obligation. Article 50 A imposes duty on each citizen to protect environment. Article 48 A, narrates that "Every State shall endeavour to protect and improve the natural environment including forests, lakes, rivers and wildlife, and have compassion for the living creatures". It, therefore, becomes the duty of each of the citizens and the State Government to protect environment with all their might and resources. Unfortunately, nobody took this aspect of life seriously till the International conventions and acts like Forest Act 1927, Wildlife Protection Act 1972, Forest Conservation Act 1980 and Environmental Protection Act 1986, have been enforced.

Impact of Environmental Pollution

A World Bank Report in 1996 has indicated that more than 40,000 peoples die prematurely every year in India because of air pollution. Delhi is the fourth top polluted city in the world and vehicular emission accounts for more than 70 per cent of the total air pollution. According to the Central Pollution Control Board Survey (1989), on vehicular pollution, the total emission of particulate matter, sulphur dioxide, oxides of nitrogen, carbon and hydrocarbons in Delhi was around 872 tonnes/day. Acid rain is another major irritant in air pollution. The emission of carbon dioxide is going up and up, resulting in the warming of earth's atmosphere. In the world, India is placed sixth in producing the largest (0.2 tonnes/capita) emission of carbon dioxide. The main source of this poisonous gas is the burning of fossil fuel (Bedi and Pattnaik, 1999). According to the study of National Institute of Oceanography (NIO) in Goa that fertilizer runoff into the Arabian sea of the West coast of India may be stimulating the production of nitrous oxide (laughing gas) and hydrogen sulphide. Nitrous oxide is a potent green house gas and some estimates suggest it is 200-300 times more potent than carbon dioxide, notorious for causing global warming. The hydrogen sulphide presence could badly affect the local fisheries industry (Bagla, 2000). The industrial pollutants in the Sutlej and Beas water are threatening the fish life to the Harike Wildlife Sanctuary. A few poachers enter the dirty water to catch fish. When they come out, they have scars on their feet and blisters on their skin. That is not enough: blood oozes from the tom muscles of calfish in their hand and there is even a bigger danger to toxins may enter human bodies through the fish. It has been warned that the migratory birds, which eat small fishes, and even the human beings, if they consume polluted fishes, would be affected by the toxins (Singh, 2001). The increased use of chemical fertilizers, pesticides and fungicides in order to achieve self sufficiency in foodgrains productions has also polluted the environment and is causing slow deterioration in soil health. The chemical residues in the food product are also causing injury to human beings and cattle population.

In rural and semi-urban sectors having less number of industrial establishments, water pollution is mostly due to contamination of water with human wastes. Recently in Jammu city 70 per cent of the samples of drinking water picked for tests were stated in unsatisfactory category and unfit for drinking purposes (Gupta, 2001). Utilization of contaminated water causes many water-borne diseases such as diarrhoea, trachoma, intestinal worms, hepatitis, cholera, typhoid etc. A World Health Organization (WHO) indicates that about 21 per cent of all communicable diseases in India (11.5 per cent of all diseases) are waterborne. These diseases kill more than 3 million peoples and cause 900 million cases of illness in the world. About 10 million infants die each year in developing countries (Bedi and Pattnaik 1999).

Poverty is a major cause and effect of global environmental problems. It is theoretically futile to attempt to deal with environmental problems without a broader perspective that encompasses the factors underlying world poverty and international equality (Alagh, 2000).

REFERENCES

Alagh, V. K. (2000). The incredible lightness of being Editorial page of *The Indian Express,* December 19, 2000.

Bagla, P. (2000). Laughing gas brings tears to West Coast. *The Indian Express.* November 16, 2000.

Bedi, D. S. and S. Pattnaik (1999). Environmental conservation in India. *Employment News.* 5-11 June, Vol. 24, pp 10.

Gupta, P. (2001). Jammu drinking water samples fail tests, found to have bacteria indicating presence of faecal matter. *The Hindustan Times,* July 11, 2001.

Singh, V. J. (2001). Pollutants and sewage wastes in Sutlej and Beas causing muscular disease among the fish in wetland. Before you bite into a Harike catfish, think twice. *The Indian Express,* May 13, 2001.

2
AIR POLLUTION

Introduction

Our planet was hitherto never confronted with so many dangers as it is facing today. Man himself is the real culprit in this regard. Increase in human population has necessitated the rapid increase in industrialisation and other activities, which have played prominent role in degrading air, water and land or soil.

Earth is surrounded by a blanket of air upto the height of about 150 km from its surface or even up to 500 km. This blanket of air is called atmosphere which covers a reservoir of 5×10^{18} m^3 of air. Pollution at hazardous level is taking place in atmosphere as well as in hydrosphere and lithosphere or soil. In fact, the entire world is in the grip of the man made pollution and all living things are now threatened out of existence.

Man made air pollution is not new, but it started as soon as man learnt to use fire. However, during the last two decades man-made air pollution has assumed menacing proportion.

Composition of Air and Definition of Air Pollution

Normally, the dry air contains a number of useful gases such as oxygen (20.94 per cent), nitrogen (78.09 per cent) and carbon dioxide (0.035 per cent) by volume in a fixed ratio. The remaining 0.94 per cent is composed of gaseous mixture of hydrogen, ozone, helium, argon, crypton, neon, nitrous oxide as well as very small amount of some organic and inorganic gaseous materials.

Definition

Basically, the air pollution is the presence of extraneous or foreign materials or substances in air beyond its normal composition. Various extraneous materials continuously enter into the atmosphere through both natural and man made processes going on the earth and cause the atmospheric pollution. These materials may be in the form of dust, fumes, smoke, gases, etc. These substances interact with the environment to cause toxicity, diseases, disorders in plants and animals including human beings.

AIR POLLUTION - WORLD SCENARIO

The world's cities which cover just 2 per cent of earth's surface, but account for 78 per cent of the carbon, nitrogen and sulphur emissions and 76 per cent of Industrial wood use (Arora, 1999). Today, atleast 1.1 billion people choke in unhealthy air according to World Watch Institute Study. Air pollution is the cause of 15,000 premature deaths in the capital of Bangladesh i.e. Dhaka every year and is the reason for the rise in chest related diseases which affect millions of the people.

Sweden blames that Industrial Plants of Britain and France are polluting the atmosphere. But Norway has charged that West Europe and Britain are responsible for 90 per cent of the acid rains in Norway. Due to acid rains, the growth of Vegetation has come down by 15, 8 and 37 per cent in Scandinavian countries, Germany and Czechoslovakia, respectively. The famous black forest of the Alps, once a pride of Europe, are seriously affected by acid rains, Acid deposition is also one of the main agents for the decline of Switzerland's forests; 43 per cent of the conifers in its Central alpine area are already dead or dying.

Every year, between 130 and 145 million metric tonnes of sulphur dioxide are released across Europe and North America and more than 90 per cent of the deposited Sulphur is man made and the rest is through volcanic eruption. Former East Germany has the highest per capita of sulphur dioxide in the world, and Poland and Slovak republic, and Hungary are among the most air polluted countries on earth. Acid rain damage is also spreading in the Third World Countries as China, India, Mexico and Brazil. The effect of acid rain would be much more widespread if nature had not provided many areas with their own protection. Alkaline soils like those covering most of Midwest US and much of South-East England, can resist acid fall out because of the presence of calcium in soils which neutralises or buffers the acids.

The incidence of air pollution from metals has reached such an alarming levels that environmentalists are finding it difficult to enforce measures for precaution and control. These metals include lead, mercury, chromium and cadmium which tend to contaminate the air around factories where they are processed. In USA and Canada, the daily intake of cadmium ranges from 50 to 80 micrograms through food chain. The pollution potential of mercury in West Germany is more than three times that leads worldwide. It has gone 4.5 times over the values recorded in 1990. Some regions of Russia are so much affected by air pollution that Krishi women suffered from skin disease caused from exposure to toxic fumes of nearby industries/factories.

Air Pollution in India

In India, 4 million tonnes of sulphur dioxide, hydrocarbons, carbon monoxide and nitric oxide are added to the air. In big cities like Kolkata, Mumbai, Chennai and Delhi, Lucknow and Kanpur the air pollution due to these gases has become so acute that it might become impossible to draw a lung full of fresh air by 2010. In greater Kolkata alone nearly 30 per cent of the students are stated to become susceptible to bronchial and other respiratory ailments due to smoke laden air caused by more than 2,50,000 vehicles plying daily.

The concentration of the above gases in the air when react with the ozone layer try to thin it or create the holes. Once the holes are formed in the ozone layer, the ultraviolet rays of the sun reach the earth and cause the skin cancer. Ozone hole was first discovered in 1979 in Antarctica (South Pole) region (Verma, 2000). Over some parts of Antarctica, upto 60 per cent of the total overhead amount of ozone is depleted during September and October. The recent observations made by Geneva based World Meteorological Organisation have revealed that during July to September, 1995 the depletion of ozone in the atmosphere over Antarctica was the most rapid ever recorded in the past. The area with severely depleted ozone now covers about 10 million km^2.

The ozone depleting compounds contain various combinations of the chemical elements like chlorine, fluorine, bromine, carbon, hydrogen and oxides of nitrogen. The compounds that contain only carbon, chlorine and fluorine are called Chloro Fluoro Carbons (CFCs). The world consumption of CFCs works out to be more than 1.2 million tonnes year^{-1}. They remain inactive in troposphere and it takes 30-40 years for these chemicals to reach stratosphere (Verma, 2000).

The Carbon deposition in the atmosphere in different forms which are also known as green house gases CO_2 (63 per cent), CFC (23 per cent), CH_4 (14 per cent) has considerably increased during recent years and it is likely to go up further (Singh *et al.*, 2000). These green houses gases are increasing at the rate of 1.5 ppm year^{-1} on the global basis. If this current rate of emission continues, this may create a green house effect, resulting in warming up of the atmosphere by 1.5 to 4.5°C by the year 2030. As the temperature will rise, the ocean will warm up, glaciers, polar ice caps may melt, increasing the flow of water into ocean, raising the sea level and flooding of arable lands in countries having vast coast line including India. Global warming also affects plant and animal species both on land and sea, which may not be able to adopt to frequent environmental changes. Due to depletion of ozone layer, photosynthetic activity in plants is hampered reducing crop yields.

A study on air pollution relating to chest morbidity in central and north-eastern Mumbai abound in textile mills and fertilizer factories indicated that human beings living in these areas were afflicted with serious respiratory systems and lung dysfunction. In Delhi, the conditions of respiratory systems are also not better. Apart from vehicular pollution, most parts of Delhi are threatened by flyash as a result the incidence of respiratory diseases in Delhi is 12 times more than the National Average (Annonymous 1995). Air pollution has greatly affected the old and the children. Children brought up under highly polluted atmosphere suffer from blood pressure, kidney and lung problems like bronchitis, asthma and acute respiratory infections (Suri, 1999).

Air samples testings from 25 selected spots in Delhi showed dangerous concentration of carbon monoxide, hydrocarbons, aldehydes, nitric oxides, and benzopyrene. Benzopyrene, a cancer causing agent is emitted by the two wheelers whose two stroke engines release almost 30 per cent of the input fuel into the air as unburnt hydrocarbons. This is why Delhi has now emerged as the fourth most polluted city in the world, first, second and third being Beijing. Kolkata and Jakarta (Kaul, 1994). As per the estimates, the level of suspended particulate matter (SPM) is at 460 Mg $m^3$of air which is much higher than the safe level of 200 prescribed by the World Health Organisation. The highest SPM level was recorded in Kanpur (470.9 μg m^3 of air) followed by Agra (423), Bangalore (90.8) and lowest in Shillong (42.2). Further it is found that on an average, a Delhite visits a doctor atleast 15 times a year for health problems, especially caused by air pollution.

The labourers working in the coal mines, sulphur mines and lead mines inhale a variety of gases and die of cardiac and pulmonary disorders *vis-a-vis* silicosis and fluorosis. In the Industrial belt of Durgapur-Jamshedpur across the West Bengal-Jharkhand border, medical practitioners have reported a phenomenal increase in stomach disorders and pulmonary ailment due to air pollution, polluted by burning of coal. The toxic chemicals used in factories for preservation of the foods also cause air pollution. The longevity of the man and the animals is decreasing due to the consumption of contaminated food. Air pollution has assumed an alarming proportions in the Ludhiana city, which faces the danger of an epidemic. In fact, respiratory and cardiac ailments have already been reported. Ludhiana city has nearly 6 lakh vehicles which are fuel inefficient. They are a major source of harmful emissions such as sulphur dioxide, carbon monoxide, hydrocarbons and lead. Air pollution is also on the rise because of the burning of rice husk and old tyres in furnaces. Punjab Pollution Control Board maintains that Ludhiana has largest population and is also the most industrialised city in the State, so air pollution is somewhat natural.

Air of north-east hills and north-west Himalayas has become poisonous these days due to influx of the tourists and frequent occurrence of land slides and earthquakes as well as forest fire. Among non-industrial pollutants asbestos exposure in Western India and silica dust in Ladakh region of Jammu and Kashmir and Lahaul-Spiti areas of Himachal Pradesh, have been reported (Singh *et al.*, 2000).

Like Delhi, in the winter capital city *i.e.* Jammu of Jammu and Kashmir State too the level of benzene and sulphur is increasing very fast, especially their levels are high after Darbar move from Srinagar (Sharma, 2000). The vehicular pollution is one of the biggest problem of the air pollution, contributing total pollution load to the extent of 2793 kg day^{-1}. The ambient air quality in several sites of Jammu city was calculated by the scientists of Pollution Control Board and found that during day time the pollution load on the air started increasing around 9 a.m. and continued till 8 p.m. About 4 years ago, the concentration of both sulphur dioxide and nitrogen oxides was within the prescribed limits but it has now crossed the norms. The concentration of SPM was found to exceed almost at all the monitoring sites due to excessive number of vehicles plying on the roads.

Sagar Kanya shows that post-monsoon area of 1,80,000 km^2 of the Arabian sea has become low in oxygen, leading to the death of bottom living marine life. In fact, over the past three or four years there has been less oxygen in this region probably as a result of increased input of nutrients from human activities. This has been accompanied by the highest accumulation of nitrous oxide and hydrogen sulphide so far observed in open coastal areas.

The magnificent historic buildings of Krakow Polland (a world heritage site), are being destroyed by acidic smog and their walls and roofs have weakened. Acid rains and other air borne pollutants are damaging the Athen's monuments. Experts say that more damage has been done in the last 25 years. Many of the Europe's stained glass windows are fading and libraries in the U.S.A. have had to be fitted with special anti-pollution equipments to preserve precious books.

Over 300 national monuments of India are disintegrating rapidly due to air pollution. Whether it is the Taj Mahal or old Jaisalmer Fort or Ajanta and Ellora caves all have started fading into oblivion due to air pollution. Air pollution caused by 1,715 industries forming a concrete belt around the Taj Mahal in Agra (U. P.) and exhaust fumes of more than 30,000 vehicles plying on the nearby National highway have turned the white marble slabs of the Taj into white yellowish and flaky.

Recently, the Central Pollution Control Board has estimated that nearly 70 per cent of the air pollution is due to the emission of two stroke engines (Aarti, 2002). A new study on pollution conducted in Kolkata reveals that besides, facing all other evils of air pollution, traffic police chronically exposed to harmful automobile exhausts also run the risk of alterations in the basic structure of their genes.

Causes or Sources of Air Pollution

The air may become polluted by a number of sources such as energy and heat generating systems which produce carbon monoxide, sulphur dioxide, oxides of nitrogen etc. Industries, specifically thermal power stations, iron and steel plants, fertilizer units, copper and aluminium smelters, oil refineries and pulp and paper making, are potent sources of air pollution. Other industries contributing to pollution are basic drugs, dyes and petrochemicals, pharmaceuticals, sugar mills, tanneries, pesticides etc. Major pollutants include carbon monoxide, carbon dioxide, sulphur dioxide, oxides of nitrogen and hydrogen sulphide. The other compounds which behave as pollutants are some aldehydes, phosgenes, detergents and heavy metals like nickel, beryllium, arsenic, tin, vanadium, lead, cadmium etc.

Smoke and Grit

Smoke and grit are produced by the domestic hearths, chimneys of the factories and automobiles (motor vehicles). Smoke is the visible product of incomplete combustion and is always accompanied by gases formed during the burning of fuels. Smoke from burning of coal consists of tiny particles with variable chemical composition, consisting primarily of carbon and some hydrocarbons. Smoke also contains chemical pollutants *viz*. 3,4, benzopyrene. Smoke and grit (atmospheric ash) are considered as physical pollutants.

Exhaust Gases

Unlike smoke, the gaseous products of fuel combustion are generally invisible. These are in the same physical condition as the air. The main exhaust gases which come out with smoke from the automobiles and chimneys of factories are : carbon dioxide, oxides of nitrogen, hydrogen sulphide and hydrogen fluoride. These exhaust gases also contain particles of heavy metals such as lead, copper, zinc and cadmium.

With the increase in population and vehicular traffic, Jammu city has earned the proud distinction of having the largest population-vehicle ratio in the world, next only to Tokyo. In 1971, it was estimated that in Jammu vehicles of all kinds numbering 10,615, contributed pollutant emission of about 30,000 tonnes annually. By the end of the year 1991, the vehicular traffic has increased from 10,615 to 85,000 which means about eight fold increase in the emission pollution. The number was further increased to 2,40,547 up to March 1998 with an annual addition of about 22,000.

It has also been observed that among all the motor vehicles, mini buses are the main contributors for vehicular emissions. The number of minibus was merely 60 in 1978 which contributed around 360 tonnes emission pollution annually but the number has increased to 1750 in 1991 exhibiting roughly 60 fold increase in vehicular emission. Similarly in Srinagar city an enormous increase in motor vehicles has occurred with no control on pollutant emission. As a result, the quality of the air has very much deteriorated in this city.

The quality of the air of other regions of Kashmir Himalayas has also been affected directly owing to the increase in vehicular traffic, emitting thereby, tremendous smoke as well as carbon monoxide and carbon dioxide in the air. The vehicular number is considerably on the rise in valley of Kashmir. The number of registered vehicles increased from 25,253 in 1986 to 80,143 in 1997. This effect has been felt more in regions having steep slopes where the rate of fuel combustion in high (Bhat, 1992). Although on the whole, the air quality does not appear to be adversely affected as yet in the valley, yet some tourist spots like Sonamarg, Kangan, Tullamulla and Gund show vulnerability to damage to air, if proper management practices are not followed to control this hazard.

Over the last two decades the tranquillity of a number of U.P. hill stations has been rudely disturbed by heavy traffic. Mussoorie, Nainital, Dehradun are worth mentioning where increasing number of cars and buses carrying tourists have contributed to the pollution. Vehicles were once not allowed to ply beyond the bus station, but now the busy Mall Road (Nainital) is clogged with cars, tourist buses and trucks, resulting in production of much more exhaust.

Similarly with three-fold increase in vehicular population over a decade in hill resorts of Himachal like Kulu Manali, Dharamsala, Palampur, Shimla pollution levels are soaring.

Industries/Factories

The rapid industrialization throughout the country has greatly damaged the ecosystem. The pollutants that are emitted through industrial chimneys are adding millions of tonnes of exhaust gases *vis-a-vis* smoke in the atmosphere. The higher concentration of these gases is responsible for the depletion of the "Ozone layer" and global warming.

Though there has been much increase in the number of factory units in Jammu and Kathua districts, the pollutant emission in this sector is slowly rising. A few units have installed pollution control systems. Yet, while the pollutant emission is marginal the pollutants are mainly in the form of effluent water and solid wastes. Their role in polluting the soil and water have been given in "water pollution" chapter.

The growth of cement factories is polluting the atmosphere unabated. The effect of cement factory dust has been seen in Kashmir in the diseased crop of saffron, fruit trees and other tree species. The effect of cement kiln on the growth of *Shorea robusta* forests around Rajban (H.P.) has been reported (Gupta *et al.,* 1992). There was decrease in chlorophyll and starch content of the leaves.

Mining and Blasting

The mountains slopes are being blasted for the exploitation of mineral resources. These mineral resources can be classified into number of categories such as metallic deposits, non-metallic deposits, precious building stone material and atomic fuels. Minerals are obtained from rocks of all types but some are found near the surface and others at depths.

While extracting the metals, non-metals and minerals, a number of mining operations *viz.* crushing and powdering, concentration of the ore, calcinations etc. are done in addition to blasting of rocks. All these mining operations emit a vast quality of dust, sand, silica and fluorine in the atmosphere. The labourers working in the coal mines, sulphur mines, lead mines and stone mines inhale a variety of gases which cause a number of respiratory diseases. People working in the sand rocks are often subject to silicosis disease and those working in the coal mines die of carbon compounds. Labourers working in the salt mines often become bald and those working in fluorine mines die due to fluorosis.

The limestone mines on different hills of Sirmour district (H.P.) are being exploited extensively and unscientifically. This has not only created a large number of furrows and channels for accelerating soil erosion, but has also caused a huge amount of dust to pollute the atmosphere. Extraction of dolomite in hill slopes of Upper Jainti basin (North-East Himalaya) also generates lot of dust thereby polluting the atmospheric air (Basu *et al.,* 1995). The polluted air is responsible for many diseases such as conjestion, mental tension, high blood pressure, mental disorder, skin diseases *vis-a-vis* silicosis in the area.

In Jammu and Kashmir, the gypsum mining in Kishtwar, bauxite mining in Reasi and lignite mining in Kashmir and Kalakote and quarrying in Ramban have also changed the environmental scenario of the State.

Biofuels

The study conducted by Indra Gandhi Institute of Development Research, Mumbai in the North-East Hills, West Bengal and Maharashtra revealed that 4 to 5.7 lakh premature deaths of children (below 5 years of age) and adult women occur due to indoor air pollution caused by biofuels. Indoor air pollution had more impact than outdoor air pollution. Further, it was found that indoor air pollution had more striking effect than Urban outdoor pollution. It is because the indoor pollutants released from unprocessed biofuels in kitchens like cow dung, fuel wood and crop residues, are far more hazardous and are 1000 times more likely to reach people's lungs than pollutants released outside.

Indiscriminate Use of Pesticides

The indiscriminate use of various weedicides for eradicating weeds, insecticides for controlling insects and fungicides for checking fungal diseases in modern agriculture, is responsible for polluting the atmosphere, water and soil. These chemicals are released into the environment as spray dusts and granules. The persistent effects of some of these chemicals and their metabolites such as malathion, DDT, BHC, is a serious threat to mankind today, so much so that in some countries the use of DDT and parathion has to be banned.

This matter is of grave concern in Jammu & Kashmir, especially in Kashmir region. In fact, very little of the applied pesticides reach the target and most of them are absorbed by the environment. Against the National average of 560 g ha^{-1} of pesticides Kashmir region uses 6,000 to 12,000 g ha^{-1} to control scab and other diseases in apple. The extensive and continuous use of the pesticides will not only have a deleterious effects on the atmosphere air but will also influence the water and soil resources. Such an effect has also been noticed in apple growing areas of Himachal Pradesh like Kulu Manali, Rampur, Kotgarh, Kot Khai and many others. Although there is not enough quantitative data regarding air pollution caused due to intensive use of pesticides, yet it is believed that if the spread of orchards and chemicals continued, the time would not be far off when these areas would suffer a severe crises from air and water pollution.

Organic Wastes, Garbages and Night Soil

The pollutants are released into air from the garbage, refuse and other organic domestic wastes. The decaying vegetation/ decomposed organic wastes impart bad odour in the air. The disposal of night soil of cities and towns is a big problem. In Srinagar the whole city is a vast latrine smelling of various gases such as methane (CH_4), hydrogen sulphide (H_2S), methyl merceptan (CH_3SH) and ethyl merceptan (C_2H_5SH). In Jammu city, the night soil is dumped into number of sites by the municipal authority which not only produces bad smell but also supplies contaminants in the air.

The rotting of horse dung all along the way from Katra (Ban Ganga) to Vaishno Devi temple, on mountain slopes creates offensive odour. Same is the case of our picnic spots like Jhagar Kotli and religious places like Chichi Mata, Peer Kho and many other places where drainage wastes give very offensive smell and one feels vomiting and giddiness. Dumping of solid wastes, garbage and drainage wastes in religious places of Himachal Pradesh like Chamunda, Kangra, Jawalaji, Chintpurni and Deot Sidh temples due to increased number of pilgrims give out a filthy odour. Similar is the situation in case of religious places of U. P. Himalayas like Rishi Kesh, Haridwar, Joshi Math and many others.

Burning of Organic Wastes

It has been estimated that in Jammu city, the quantity of refuse generated per capita per day varies from 115 to 500 g. The refuse mainly consists of paper wastes, glass, plastics, ash, fine earth, cotton/nylon rags, fruit peels, leather left over, covers of medicine etc. Open burning of this solid waste is a common practice in northern Indian cities including Jammu. The burning of the solid waste is generally done either in community bins or near the trunk of living tree planted on the road side. Burning of this waste not only creates air pollution, fire hazards and sometimes destruction of useful trees but is also waste of useful resources which could be recycled.

Radioactive Elements

Radioactive elements from various sources constantly pose an environmental risk. These are mainly given out as fission byproducts by atomic explosions, testing of nuclear weapons refrigerators and atomic power reactors. The radioactive dust that falls to the earth after atomic explosions is called radioactive fallout. The effect of the radioactive elements present in radioactive fallout depends upon their half life periods. Longer the half life period of radioactive element more potentially dangerous is the isotope to man and animals. For example calcium-45 has half life period of 160 days whereas strontium-90 and cesium-137 have still longer half life periods.

Ionizing radiations

Ionizing radiations given out by atomic explosion and testing of atomic weapons include alpha and beta radiations, neutrons gama rays etc. Their effect was realised after the explosion of atomic bombs on the cities of Hiroshima and Nagasaki in Japan during second world war. After the explosion these cities assumed the shape of desolate land with dead bodies of animals and men scattered like the splinters of the bombs.

Hazards of Air Pollution

Human Health

Clean and pure air is very essential for survival of man and other living things. But now the air which we breathe contains many pollutants. However, the forests help in purification of air which is further corroborated by the slogan of the Chipko movement :

> "What do the forests bear
> Soil, water and pure air,
> Soil, water and pure air,
> Are the basis of life".

This slogan is relevant to whole of the world and can save mankind which is on the brink of disaster due to industrialisation.

Jammu city was unique in its sylvian beauty till 1950. It was then surrounded by a massive green belt comprising famous "Rakhs" of Ramnagar, Ploura, Roulki, Kheri etc. It was very famous for its cool breeze during the reign of Dogra rulers, locally known as "Tadu". But now we are living to languish in heavily polluted air and practically treeless city. The deforestation of Jammu city had raised the level of health hazards and has lowered the quality of life.

As already mentioned the most dangerous gaseous pollutants of the air are particulate matter, smoke, carbon monoxide, sulphur dioxide and oxides of nitrogen.

The effect of particulate matter depends upon the fate of the particles in our body. If their size is more than two micro metres they are trapped in the hairs of nasal passage or bronchial mucus and coughed out or swallowed. Small particles enter the alveoli where they may be engulfed by the special cells or absorbed in the blood stream.

Carbon monoxide when enters the blood stream, replaces the

oxygen of the oxy-haemoglobin, causing thereby serious disorder in body's metabolism. Similarly sulphur dioxide and nitrogen oxide diffuse into the blood stream. They combine with haemoglobin and impede the oxygen transport. Nitrogen oxide at high concentration impairs the functioning of lungs by causing water accumulation in the air spaces.

A large number of pollutants, especially hydrocarbons are reported to induce cancer. Heavy dust fall is one of the main causes of the prevalence of asthma, cough, bronchitis and lung cancer particularly in children. Heavy metals in the atmosphere like lead can damage the brain of young children and may even cause death. Cadmium is a respiratory poison and may contribute to high blood pressure. Lead causes gastro-intestinal disease blood pressure and number of heart diseases and damages liver and kidney.

The presence of toxic trace elements present in coal mine dusts is a serious environmental pollution. These elements especially Cu, Pb, Ni, Cd, V, U, Th, Se, T, and Cs can cause serious diseases besides already known ailments like bronchitis, asthma and silicosis. Fluoride causes a decaying of bones.

Photochemical smog is another pollutant which is highly dangerous to all age groups and pregnant women. Affects grapes, apples, cotton etc. Higher concentration of sulphur dioxide in the air not only leads to chronic type of injury in fruit tree but has deleterious effects on forest trees also.

The presence of fluorides in the air destroys the leaf tissues, fluorides and oxides of nitrogen reduce the crop yields. Ozone harms vegetables, field crops, fruits and forest trees.

Photochemical smog bleaches and blazes foliage of important leafy plants. Hydrocarbons such as ethylene cause premature leaf fall, flower bud shedding, curling of petals and discolouration of sepals.

Growth of lichens on trees are inhibited by air pollution. Therefore, lichens can serve as indicators of air pollution. Clean air is indicated by abundant growth of lichen. In U. S. A. air pollution has resulted in 5 per cent fall in agricultural production.

Animals Health

Air pollution also causes widespread damage to various animals. The general effects of air pollution on domestic animals living in or near industrial areas are almost similar to those found on human beings. Livestock ingest various fluorine compounds falling on forage crops and suffer from fluorosis disease. In this disease there is abnormal calcification of bones and teeth in animals. This also results in lameness, loss of weight and frequent diarrhoea.

Damage to Vegetation

Air pollution has been known to have an adverse effect on plants. At first, it was only sulphur dioxide that was considered a dangerous pollutant. Now, with the advent of various pesticides and new industrial processes, the range of harmful pollutants having effects on vegetation has been multiplied tremendously. Sometimes, vegetation over 150 km away from the source of the pollutant has been found to be affected. Economically important plants including fruit trees worth billions of rupees are damaged in the developed countries annually.

Economic Effects of Air Pollution

Air pollution damage to property is a very important economic aspect of air pollution. In the United States of America, Great Britain, France and many other European countries, this problem has been investigated in detail. This damage to property covers a wide range such as corrosion of metals, soiling and eroding of building surfaces, fading of dyed materials, rubber racks as well as works of art.

A large number of historical monuments are affected by air pollution. Deep concern was expressed about the possible damage to the famous marble structure Taj Mahal at Agra.

The National Ambient Air Quality Standards (NAAQS) for India are given in Table 2.1as under.

It has been actually found that pollution level is three to four times more as compared to above NAAQs data.

Table 2.1 : 8 hours concentrations in $\mu g\ m^3$.

Category	SPM	SO_2	CO	NO_2
Industrial & mixed	500	120	5000	120
Residential and Rural	200	80	2000	80
Archaeological & Forest sites	100	30	1000	30

The consumption and wastage data of 1 million people in city on the average is given in Table 2.2 as under :

Table 2.2 : Consumption and Wastage Data

Material	Consumption (Tonnes day^{-1})	Wastage (Tonnes day^{-1})
Water	5,25,000	5,20,000
Food	20,000	2,000
Fuel	9,600	950
	year $^{-1}$	tonnes as air pollutants

Effect of Increased Level of Carbon Dioxide in the Atmosphere

The geologists believe that when the earth was formed about 4.5 billion years ago, the atmosphere did not have any oxygen. It was mainly composed of carbon dioxide, water vapour and methane gas only. Oxygen gas appeared in the atmosphere only after green plants evolved the photosynthetic process to convert the carbon dioxide and water vapour into carbohydrates and release oxygen. As green plants proliferated the carbon dioxide level in the atmosphere gradually dropped. Over millions of years, the atmospheric carbon dioxide level stabilized at around 290 ppm before the Industrial Revolution of the late nineteenth century. But since then, the level of carbon dioxide has been increasing steadily because of large scale burning of fossil fuels and wanton deforestation and today it stands at a little over 350 ppm and is likely to rise to 440 ppm in next 50 years.

Green House Effect

Although the amount of carbon dioxide present in atmosphere is relatively very small, the effect it has on the terrestrial temperature is significant. Carbon dioxide acts like glass windows in a green house allowing warm rays of sun to pass in, but holding back at the same time, the heat radiated by the earth. In this way it traps and reinforces the solar heating effect and causes, what is popularly known as the "Greenhouse effect".

Climatological Changes

Scientists affirm that the carbon dioxide content in our atmosphere rises by nearly 30 per cent every century, which in turn raises the earth's temperature by 1°C. As early as 1861, the English Philosopher, John Tyndall proposed that increased concentration of atmospheric carbon dioxide could conceivably raise surface air temperature and produce a change of climate similar to that created by green house. Since, then in the course of century, the concentration of this gas has increased from 280 ppm to 350 ppm an increase of about 25 per cent. This increase has been caused by various factors such as indiscriminate burning of fossil fuels like coal, oil and natural gas, aviation fuel and deforestation etc.

According to one estimate, the average global temperature may rise by as much as 5 degree Celsius by AD 2100, which could have disastrous results. One effect of such warming would be rise in the sea level upto 200 cm which would inundate large coastal areas throughout the world.

The atmospheric warming would also affect global weather pattern. Computer modelling has shown that such warming may turn large areas of fertile land into deserts by disruption of rainfall pattern, while other areas which now get little rain may get torrential rains.

Disruption of Ozone Layer

Another crucial atmospheric phenomenon that has been worrying environmental scientists is the discovery of a ozone hole over Antarctica. It is known since long that there is a layer of ozone gas in stratosphere. Ozone is produced in stratosphere by the action of higher energy ultraviolet radiation on oxygen, and being a good absorber of ultraviolet radiations its presence there effectively blocks out the sun's ultraviolet radiation. Ultraviolet radiations are dangerous to any life form and cause skin cancer in human beings. The stratosphere ozone layer thus plays an important protective role for life on earth and its disruption is obviously fraught with dangerous consequences. It has been estimated that one per cent drop in ozone level in the stratosphere could lead to as many as 10,000 more cases of skin cancer a year in the United States alone.

A group of man made compounds responsible for destruction of ozone layer of the atmosphere, are chlorofluoro carbons (CFCs). These compounds are being extensively used in refrigeration industry. Bromofluorocarbons (BFCs) are the other compounds that deplete the ozone layer. These are released in the atmosphere from agricultural practices using methyl bromide as insecticide.

Effect of Higher Concentration of Sulphur Dioxide and Nitric Oxide

When the concentration of sulphur dioxide and nitric oxide increases in the atmosphere, they cause acid rains in the form of sulphuric acid and nitric acid. This phenomenon is much more prevalent in advanced countries notably the USA, UK and other European countries on account of large scale industrial activities. Acid rains have destroyed black forests of Germany. These rains have also reduced the forest growth in Scandinavia. In Canada, thousands of lakes have been destroyed due to acid rains.

Black Snow/Black Rains

The occurrence of the black snow in the northern mountain ranges of Kashmir was seen during 15th March, 1991. It was suggestive of westerly disturbances originating from Mediterranean sea, sweeping the smoke from Gulf region arising from the burning wells of Kuwait during Iraq-USA Iran war. These westerly disturbances with smoke passed through Iran and Baluchistan and finally precipitated in the form of Black snow in the Gund area (Sonamarg) of Kashmir and in the valley as greyish to black rains. Black rains due to fallout of Gulf war were also seen in Jammu area on 2nd of April, 1991.

The analysis of black rain water samples indicated that water so collected contained suspended black lumps in colloidal state and it was oily in appearance. The samples showed presence of burnt hydrocarbons and sulphur. The thin section studies of the residues after filtration of water samples, revealed carbon complex in the form of lumps and quartz grains in minute euhedral shape.

The filtrates of black rains were analysed in respect of heavy metals (Dhar *et al.*, 1991). A comparison of the concentration levels of the elements viz. Ni, Ga, Na, Zn, Mo, Fe, V and Fb in normal rain samples during the period 1984-1990 with of black rain in 1991, shows a substantial increase of Cu and Ga in the latter.

There being no major chemical or other heavy industry in and around Jammu, the appearance of black rain cannot be construed as a local effect. The only plausible explanation for its incidence seems to be the contamination of the atmosphere due to massive burning of petroleum in Gulf region.

Control Measures

To prevent further deterioration of the air pollution certain urgent actions have been highlighted by a number of renowned environmentalists. Fortunately, some of these urgent actions discussed below are workable solutions to this man made hazards.

Plantation

Intensive plantation in the region considerably helps in reducing dust, smoke and other pollutants, People should be made aware of role of trees and forests in air pollutions control. Urban forestry and rural forestry programmes should be encouraged while planting the trees.

Rural Forestry

It is also called Extension Forestry. Rural forestry encompasses forestry on community and panchayat lands, degraded forests, waste lands, roads and railway sides and canal banks etc. for meeting the needs of the rural people. It also includes restoration of degraded areas bearing scars of quarrying, mining, road, constructions lime burning and brick kilns.

The studies have shown that leaves of trees help to control dust pollution which comprises around 40 per cent of the total air pollution problems in India and also the level of toxic elements in atmosphere trapping them on the leaf surface.

It has been found that trees with simple broad leaves i.e. Pipal (*Ficus religiosa*), Borh or banyan (*Ficus bengalensis*), Teak (*Tectona grandis*) Sal (*Shorea robusta*), Mango (*Mangifera indica*) and Kachnar (*Bauhinia variegata*) etc. are more effective in removing dust particles suspended in the air than those having compound and smaller leaves like Ashoka (*Saraca indica*), Neem (*Azadirachta indica*) and Gulmohar (*Delonix regia*). Therefore, along the road side and at places where maximum dust generation is possible, trees of simple leaf should preferably be planted.

Urban Forestry

Urban forestry aims at protection and preservation of existing trees and growing new trees in vacant spaces, along roads, in private and government compounds etc. in order to enhance natural environments of the city and towns.

Green belts should also be raised in urban and industrial areas as well as in cold arid areas of the States. Such programme will also check soil erosion, desertification as well as improve the microclimate.

Ventilation/Exhaust Fans

Indoor air pollution can be controlled by proper ventilation, avoiding biomass fuel and using gas for cooking. Exhaust is a dire necessity in the kitchen. Avoid attached bath rooms, tobacco smoking in confined space, and keeping the house free from dust and dirt.

Restriction to use certain types of Fuels/Use of certain well Run Public Transport

In a relatively more polluted and high traffic intensity region, pollution due to automobiles can be controlled by restricting the use of certain types of fuels. For example, in extreme cases of air pollution, a complete ban on diesel driven vehicles can be imposed within certain areas.

Air pollution caused by the automobiles can also be easily countered by adopting a series of measures like well run public transport, using bicycles, encouraging to walk on foot, avoiding frequent use of cars, using lead-free petrol, regular servicing of vehicles, sticking to speed limits and fitting the vehicles with catalysers to minimise output of carbon monoxide.

Prohibiting the Use of Unapproved Appliances

Control of air pollution from domestic, commercial and other sources including very small industries can be possible by prohibiting the use of unapproved appliances. At best, on residential areas ban on use of hard coke can be imposed on the use of a particular type of *Chulas* or ovens to ensure control of smoke from domestic fires can be insisted upon. Similarly pressing problems of energy crises in industries can be solved effectively by decreasing fossil fuel consumption and making use of other energy sources like sun, wind, rivers, tidal waves etc. The decrease in fossil fuel consumption will help in minimising air pollution.

Air Cleaning

Air cleaning device removes or recovers suspended solids, liquid and also soluble gases and vapours from air before they are discharged into the atmosphere. The commonly used air cleaning devices are :

(a) Gravity settlers,
(b) Cyclone separators,
(c) Wet collectors, and
(d) Electrostatic precipitators.

Physical Methods

Physical methods include the suitable height and design of the stack of factories. The minimum height of the stack has been worked out to be 38 metres in order to ensure proper mixing of the discharged pollutants into the adjoining atmosphere. In fact, by this method the effect of pollutants gets diluted. So this method is also called dilution method.

Chemical Methods

These methods are used for removing sulphur dioxide from air, resulting from coal burning with rich sulphur content. Several chemical methods have by now been developed to have an effective control over air pollution problems. Briefly some of these methods are discussed below :

Beneficiation Process

By this process coal is treated with lime or dolomite and processed in such a way that both sulphur and ash control of the original coal are significantly reduced with more energy or non-burning.

Reduction Process

In this method, the sulphur dioxide emission is absorbed in alkaline alumina solution, resulting in quick reduction of the former into elemental sulphur.

Catalytic Oxidation Process

In this method, the vanadium is used as a catalyst, whereby sulphur dioxide is converted into sulphuric acid or ammonium sulphate.

Inceneration, Adsorption, Absorption, Condensation

These techniques are used to control hydrocarbon emission from stationary sources. In adsorption process, exhaust gases are passed through a bed of granulated absorber where hydro carbon vapours are adsorbed into the absorber. The absorption method is similar to adsorption except that the exhaust is brought into close contact with a liquid in which the hydrocarbons dissolve or become suspended. In condensation method, exhaust gases are passed over low temperature surface where they are condensed. Liquid hydrocarbons are left behind are collected.

Priorities and Suggestions for Future Improvement

To control air pollution, the following priorities and suggestions are proposed:

(i) To educate the citizens and rural masses about environmental risks, economic and health hazards, is the need of the hour, which must be given top priority. For this purpose, conferences, seminars and symposia should be arranged by the Government institutes and private organisations. Information regarding environmental pollution must be published in local language also.

(ii) Assistance from government side should be provided to small scale industrial units, particularly those located in rural areas to aid the implementation of pollution control measures. This can be achieved by promoting development and adoption of cleaner technologies including friendly biotechnology.

(iii) The reckless use of pesticides should be avoided as these are deadly to all organism including the one which we intend to protect.

(iv) Integrated steps should be taken to strengthen the government and institutional agencies dealing with environmental management. The ministries dealing with the sectors of energy, industries, water resources, soil and forest resources, agriculture and others should develop specific programmes regarding air pollution prevention.

(v) An integrated over view and organisational set-up for decentralised environment impact assessment and environmental law enforcement based on cooperation with local authorities must be sought.

(vi) The collection and integration of environmental, economic and health data should be done to determine the status and to develop a concise set up of environmental indicators for monitoring the effects of air pollution.

(vii) The present standards based on the concentration of pollutants in effluents and in emissions must be revised. The norms should be revised to lay down mass based standards, which will be specific limits to encourage the minimisation of waste, promote recycling and refuse of materials, as well as conservation of natural resources *i.e.* air, water and land.

(viii) To act against potential problems in the future, new units should be conformed to stricter standards. They must select technologies that produce no or low quantities of wastes and also recycle or refuse water products. Progressively, more strict vehicle emission standards have to be evolved to deal with environmental hazards caused by vehicular traffic.

(ix) Make your environment smoke free. Do not encourage smoking at home or at work. Remove ash trays and be firm and adament. Let yourself and others live in a smoke free life.

(x) Always keep your area neat and clean. Keep your surroundings in good sanitation and fit for living. Do not throw garbage on the streets.

Impact Points to Remember

- Today air pollution is global based and it is more dangerous than aids or cancer.
- The pollution levels in all the cities of India are much higher than the safety levels.
- Polluted air in 36 Indian cities killed 52,000 people in 1995, a 28 per cent increase from 1990. The number is much higher than this figure. Air pollution is caused by industrial gases and effluents, garbages and miscellaneous wastes generated by various industries and factories.
- The mining operations and blasting of rocks for building roads for constructing factories and hydro-electric projects emit a vast quantity of dust, sand, silica and fluorine in air, causing lot of air pollution.
- Bad condition of roads, encroachments, unauthorised parking violation of traffic rules, and the absence of traffic marks besides the ever increasing pressure of vehicles are the other reasons which bring air pollution in many of Indian towns and cities.
- The regular spraying of insecticides, weedicides, fungicides, and bactericides not only pollute the air but water also.
- The radioactive contamination from nuclear plants installed all over the world including India has increased at the danger level

which has greatly affected the terrestrial, marine and atmospheric life.

- Besides human beings and animals, air pollution has started influencing plants and monuments.
- Stomach disorder, pulmonary or respiratory ailments cardiac disorders silicosis, fluorosis, cancer and many others are due to inhaling of contaminated air.
- Trees and green patches or covers have a very good effect in absorbing pollution from industrial wastes, fertilizers, factories, vehicles and other sources. It, therefore, becomes imperative to increase green cover along the roads and around the industrial areas, mega cities, cities and towns.
- While planting various tree species in agro-forestry and social forestry systems emphasis should be given to grow local species.
- In the housing colonies, small gardens, kitchen gardening should be developed. The right way to plan garden, is to draw up a list of trees which can absorb air pollutants, dust and also prevent noise pollution.
- In this regard, *Saraca indica, Bauhinia variegata, Cassia fistula, Populus alba* should be preferred.
- Some of the fruit trees recommended for the purpose are: *Emblica Officinale, Mangifera indica, Citrus aurantifolia, Citrus reticulata.*
- Industrial estates and clusters of small scale industries should be scattered throughout the countryside rather than to concentrate on one particular area.
- Success in controlling pollution can only be achieved through public participation and cooperation.
- Attempts to control air pollution at the source can be made in two ways: (i) by separating the pollutants from the harmless gases and (ii) by converting the pollutants to innocuous products before their release in the atmosphere.
- Another method of conversion of pollutants is oxidation in air. For example nitric oxide can be converted into nitrogen dioxide by oxidation.

REFERENCES

Anoymous, (1995). Status Report. Department of Environmental Science, Government of Delhi.

Arora, V. (1999). Air pollution major killer in cities. *The Himalayan Mail*, Wednesday June 23rd, 1999 issue.

Aarti. (2002). *Daily Excelsior*, 22nd April, 2002 issue.

Basu, S.R.; Dhara; N.; Sengupta, N. and Sinha, M. (1995). The effect of mining on environment - A case study in the Jainti Basin of Eastern Himalaya. In *Ecodegradation of Himalayas - Causes and Remedies* (Ed. by B. L. Kaul). Vinod Publishers & Distributors, Pacca Danga, Jammu, J&K India.

Bhat, M.Y. (1992). Ecological Implications of Tourism in the Sindh Valley (The Kashmir Himalayas). In *Himalayan Environment Man and Economic Activities* (Ed. by J. L. Raina), Pointer Publishers, Jaipur, Rajasthan, India.

Dhar, M.L.; Kalsotra, B.L.; Durani, S. and Fotedar, B.K. (1991). *Black Rain in Jammu,* Department of Chemistry & Department of Geology. University of Jammu, Jammu.

Gupta, R.D.; Kanwar, B.S.; and Rana, K.S. (1992). Ecological considerations of the Sirmour Himalayas. In *Himalayan Environment Man and Economic Activities* (Ed. by J. L. Raina), Pointer Publishers, Jaipur, Rajasthan, India.

Kaul, M. K. (1994). World Environment Day - Our deteriorating environment. *The Daily Excelsior Sunday Magazine*, June 5th, 1994 issue.

Sharma, G. (1999). Pollution gives rise to acid rains in Jammu. *Daily Excelsior Sunday Magazine*, November 7th, 1999 issue.

Sharma, G. (2000). Growing Vehicular Pollution dragging Jammuites to death trap. *Daily Excelsior Sunday Magazine* October 22nd 2000 issue.

Singh, K.K.; Sharma, P.D.; and Sarolia, S.K. (2000). Environmental Pollution in Indus Belt. *Environment and People*. 6 (11) : 6-8.

Singh, K.K.; Patrin, M. L.; and Sharma, H.C. (2000). Environmental Protection and sustainable agriculture. *Intensive Agri*. 38 (1-2) : 12-16.

Suri, B. K. (1999). Increasing Vehicular Pollution. *Daily Excelsior* Thursday August 5th, 1999 issue.

Verma, S. S. (2000). Ozone Layer Depletion : A Summary. *Environment and People*. 7(7) : 16-19.

3
WATER POLLUTION

Introduction

Water occurs in liquid, solid and gaseous state. It is a major component of environment which covers nearly 70 per cent of earth's surface. Water is very essential both for animal and plant life. Its various resources such as sea, rivers, streams, lakes and springs have always been of special interest to man owing to their multipurpose uses. Man requires water for drinking, bathing, washing and irrigation, livestocks, air conditioning and flushing of toilets and other municipal uses etc.

Apart from above, a substantial quantity of water is utilized in various industries and factories *vis-a-vis* in fighting fire in urban areas. This great necessity for water has brought into focus the fragility of environment and needs to guard it. All great civilization has grown around water and many had perished with dwindling of water resources perhaps not having realised the significance in its sustainable development.

"Water is life when pure, but becomes poison when gets polluted". Water pollution is a phenomenon that is characterised by the deterioration of the quality of land water (rivers, lakes, ground water) or sea water as a result of human activities directly or indirectly. Water is relatively more prone to pollution than air because it is more viscous, having lesser moment of inertia and possessing closer molecular distance than air. Air being available in the environment in much larger proportion as compared to water, the pollutions in the air are quickly diluted and dispersed away along the down wind direction. Pollution load is thus, drastically reduced in air. The unequal distribution of water on the surface of earth and fast declining availability of fresh useable water per capita, per year, due to increasing population, are the major causes for concern in terms of the quantity and quality of water. Thus, water pollution is more dangerous pollution.

In this chapter, first of all, shortage of fresh water in the world and water crisis in India, various human activities that cause water pollution, have been classified and, thereafter, the hazards of water pollution and their remedial measures have been discussed.

Shortage of Fresh Water in the World

Water is another boon's nature to all living beings after air but it is so much polluted today that we are reaching a stage when the water effects are globe based. Almost half of the rivers and lakes of the world are so much polluted that they no longer can be processed for drinking purpose. Presently only 3-6 per cent of the world's water is fresh and fit for drinking (Subba Rao, 2001). Shortage of fresh water is due to over population and poor sanitation *vis-a-vis* the phenomenal advance in industrialisation, agricultural and mining operations, domestic activities and illiteracy. Worldwide water use has increased six to eight fold since 1900, but the global supply of fresh water is not expanding. Significant increasing population in some of the most water stressed parts or the world can be seen from the five river basins *viz.* the Nile, the Jordan, the Tigris-Eupharates, the Ganges and the Aral sea where the projected population increase would be from 45 to 78 per cent between 2000 to 2025.

In East Germany (now the United German) out of 75,565 villages about 1000 have drinking water so high in nitrates that infants are routinely given bottle water. The mothers in the city of Berlin are worried as drinking water suffers from *Pseudopodia* due to contamination with organic wastes from the sewerage. The children cannot withstand it and die. About 35-48 per cent of the people dwelling in Sub Saharan Africa and Asia have no access to safe water. The regions of the highest concern are the Middle East and Western Asia, Northern-Eastern and Southern Africa as well as parts of Caribbean. About 20 countries in Africa and West Asia were facing acute shortage of fresh water during the close of twentieth century and now the number of such countries has become double. In African villages, the mothers offer their children water to drink despite knowing that it can cause diseases, but there is simply no other water they can use. In Czechoslovakia about 70 per cent of the rivers are polluted. Similarly many regions of Russia, are facing an acute shortage of fresh water due to water pollution. Today there are hardly any streams and rivers in Canada, except in a few high mountain reaches, that can satisfy human thirst without chemical treatment (Mohammed, 2000). The pollution of large rivers such as Rhin and Danube which pass through many countries, bear the swiss wastes to Netherlands and the Austrian wastes are carried to Hungary and Romania. (Karunakaran, 2000).

The main water pollutants include chemicals, pesticides, fertilizers, industrial and other organic wastes, toxic metals, inorganic wastes and undesirable micro-organisms. These pollutants adversely alter the quality of water, disturb and destroy the ecological balance and eventually create health hazards.

World Health Organisation has reported that more than 20 per cent of deadly infections including cholera and typhoid are carried by water. Polluted water also serves as a vehicle for the spread of giardia and cryptosporium. Other water borne diseases like diarrhoea and jaundice are the result of contaminated and polluted water.

Fresh Water Crises in India

Water famine is looming larger, say experts and warn that unless a countrywide people's movement is set in motion to recharge the depleting water table, it would not solve the water scarcity problem. Recharging ground water through good rain water harvesting techniques is possible, through rain water harvesting technology. Rain water harvesting is, indeed, catching up in Chennai, Faridabad and Kanpur, where it is a part of the State policy. In Delhi, it is compulsory to put up ground water recharging structures in areas where the water level has gone very low. Revival and new construction of ponds must receive top priority for meeting the water needs of the people and the land.

A fairly good idea of water pollution facing in India can be had by the fact that even at this stage of industrialisation 70 per cent of the rivers are already polluted with industrial wastes. Another startling fact is that out of 319 cities and towns only a few have sewerage treatment facilities while others have made shift arrangements. Two thirds of India's available fresh water is lost every year due to runoff losses to the sea and evaporation. The fresh water crises is already evident in many parts of the country like Gujarat, Andhra Pradesh, Madhya Pradesh, Rajasthan and North East Hill (States like Tripura, Manipur, Assam).

Over 25 million people from 15 States viz. Rajasthan, Tamilnadu, Maharashtra, Bihar, Delhi, Jammu and Kashmir, Uttar Pradesh, Madhya Pradesh, Gujarat, Punjab, Harayana, Karnataka, Andhra Pradesh, Kerala and Orissa, are affected due to excessive fluoride content, 2-20 ppm in potable water sources (Sharma *et al.*, 2000). As per WHO guidelines (1984), the permissible limit for fluoride in potable water is fixed 1.5 ppm. Fluoride ions have both negative and positive impact on human health. It is beneficial for human health if it is taken in controlled quantity (0.5 to 1.5 ppm). But it causes dental and skeletal fluorosis when fluoride content is more than 1.5 ppm in drinking water. This bone disease is prevalent in many parts of Assam, Andhra Pradesh, Rajasthan, Punjab, U.P., Haryana. Fluorosis in livestock is due to the excessive intake of fluoride livestock through water.

Fluorosis problem was first surfaced in the West Bengal in the late nineteen hundred eighties when in a bid to control malaria, the Government sank hand pumps so that surface water bodies could be sprayed with pesticides. As it happened the malaria epidemic was brought down but thousands fell victims to arsenic poisoning caused by the water they were drinking. Subsequently this problem was noticed in Orissa. Arsenic problem has been explained due to indiscriminate withdrawal of water for agriculture which causes the intrusion of saline water, especially in coastal areas into the subterranean reservoirs. It could also be stated due to the geological composition of the Indo-Gangetic plains. Now in 8 districts of West Bengal, more than 5 million people are exposed to arsenic menace, out of them nearly 30 per cent are children. High concentration of arsenic in ground water causes anaemia, skin cancer, diarrhoea and lung diseases. Cardiovascular damages, gastrointestinal ailments, hepatic and neurological effects are the other diseases caused by arsenic polluted water.

In many parts of the country especially the eastern region, underground strata contain soluble iron and manganese. These elements pollute ground water and hence well water in such areas have a reddish colour and metallic taste. Both iron and manganese are harmful whether the water is meant for domestic or industrial use. In some other areas the underground and surface water is brackish or saltish so desalination of this water is of great importance in places where there is acute scarcity of fresh water. Nearly half of Kolkata's population runs the risk of developing cardiovascular diseases due to high carbonate content in drinking water. Residents of Salt Lake, north and south Kolkata have been drinking brackish water containing carbonate daily for the last 25 years which has given rise to heart ailments and artery. In northern and eastern parts of Kolkata the ground water has shown chlorides exceeding 500 mg per litre of water in the form of calcium and magnesium chloride. Further calcium and magnesium bicarbonate type of water has been found in south western Kolkata. Certain isolated pockets of sodium chloride type of brackish water occurring in southern part of the city including Garfa, Dhakuria, Ranikuth and Jadavpur make this water risky for drinking.

River Water Pollution

The worst river water pollution is found in both rich and poor countries, according to World Commission on Water (WCW), which held its second global ministerial conference during March 2000 in Dutch city. In America, the most polluted rivers include Colorado which flows in the south western United States and touches north-western Mexico, and the Lerma river, which moves through middle of Mexico and finally falls into the Chapala lake. In Eurasia, the sickest rivers labelled very unhealthy by the commission consist of Amu Darya and Syr Darya, which flow into Aral sea in Central Asia, the Ganges which flows from the Himalayas to the Bay of Bengal, and the Yellow river which courses through China's most important agricultural region. Other rivers posing serious pollution problems, are Rio-Grande between United States and Mexico, the US Mississippi, South America's Parana, and three rivers in Europe, the Danube, the Rhine and the Volga.

Only healthy rivers are the Amazon in South America and the Congo in Sub Saharan Africa-both of which remain relatively untouched by Industrial activity. Also the Menkong river of South East Asia and the St. Lawrence, are better than other rivers.

In India, river water pollution is on the increase near big cities and towns. In 1969 the river Yamuna near Okhla in Delhi was largely contaminated and due to this hundreds of fishes died and floated on the water surface of the river for several days.

Today the Yamuna river is India's one of the most sewage polluted river. The Central Board for the Prevention and Control of water Pollution estimates that the Yamuna receives approximately 150 tonnes of organic waste, 150 tonnes of suspended solids and 300 tonnes of dissolved solids per day in Delhi alone. The *E. coli* bacterial count was 75 per 100 ml of water at Wazirabad but this count increased to stupendous 240,000 per m*l* of water at Okhla. A study conducted by the Central Pollution Control Board indicates that during 1982-1986, the four cities *viz*. Delhi, Agra, Mathura and Allahabad contributed 1067, 70, 51 and 155 million litres waste water per day which now has become almost double. The river water of Yumuna is, thus, not even fit for irrigation, but not to talk for drinking. It is also a breeding ground of water borne diseases like cholera, gastroenteritis, typhoid and yellow fever.

Another source of contamination of the Yamuna water, is due to the lack of proper latrine facilities for about 5 million slum dwellers, who use open drains for defecating-all these run straight into the river, carrying all their filth.

Water of the river Ganga that takes its birth at Gangotri was thought to be unpolluted upto Haridwar. But presently it is not the case, even at Haridwar it is being polluted by the discharges of number of industrial wastes. At Farukhabad water of the river Ganga becomes red due to discharge of industrial effluents. These effluents have toxic metals such as Cr, Pb, As, Cu, Co, and Ni. At Kanpur, the river is worst affected by organic and inorganic chemicals from 55 tanneries, 15 textile mills and 2 jute mills, and number of other factories. At Allahabad, fertilizer Industrial Complex and other industries discharge about 0.75 million litres of effluents in this river. The domestic sewage and effluents from printing and metallic industries in Varanasi also pollute this river. The domestic sewage and effluents from printing and metallic industries in Varanasi infact pollute this river to such an extent that it becomes even unfit for bathing.

Besides, the industrial wastes of 453 polluting units of Uttar Pradesh, Bihar and West Bengal, more than 873 million litres of domestic waste also flows into the river Ganga everyday.

The Ganga Action Plan has estimated that about 260 million litres a day of industrial effluents, including toxic wastes are dumped into the river waters making the Ganga, a highly polluted river.

Discharge of human excreta from the Howrah station railway platforms, into the river is another source of pollution of the river Ganga. The once mighty Ganga river at Patna has not only been reduced to a small stream but is also very much polluted. The bed of the holy river is fast drying up. It has, infact, moved about 500 metres from the original river front. The Ganga shows signs of chronic illness over 400 km stretch between Buxar and Bhagalpur as it passes through Bihar. Munger is one of the district which is threatened by drinking water crisis.

The quality of water of the river Betwa, a lifeline of the thousands of Vidisha and Raisen district in Madhya Pradesh is not even suitable for irrigation, let alone drinking. The report containing observations of Nagpur based National Environment Engineering Research Institute states low pH, high conductivity and high concentration of Total Dissolved Solids (TDS), Biological Oxygen Demand (BOD) and sulphates of water samples of Betwa river. The water quality of Betwa river is far inferior to drinking requirement as indicated by physico-chemical characteristics.

The river Damodar, which is an integral part of the ecological system of eastern India, has turned into a "biological desert", as a result of massive water pollution. The water of the Damodar river is considered unsafe for human use even after filtration. Besides, the river virtually remains dead due to very low level of water caused by silting and dumping of garbage as well as industrial waste. Massive coal mining, thermal generation, urbanisation and mushrooming growth of industries have been identified as the main reasons for the pollution of Damodar river. Due to the massive toxicity of river, the aquatic flora and fauna have started vanishing. At many places the river presents the appearance of a big drain carrying black, highly turbid water.

So high is the pollution that according to Bihar State Pollution Control Board, the total suspended matter at many places is 40 to 50 times above the permissible limit.The most dangerous Damodar's pollutant is coal mining. Large open cast mines, which account 60 per cent of coal production in the country are the most serious degraders of soil and pollutors of Damodar. The mechanical extraction of coal is amenable for generating waste, as the machines mine rock as well as coal. The former is discarded outside the mine. When rains come, it gets carried into the river. The other pollutor is coal dust. Around one million tonnes of this has accumulated close to Dugdha Washery in Dhanbad district in last 20 years alone. During monsoons, the rain water carries and dumps it into the river Damodar. Sewerage water of 7 lakh people of Kota is allowed to flow freely into the Chambal river. As a result, this river becomes dirtiest like the rivers Damodar and Rewa in the country.

The river Beas, lifeline of several important towns and villages in Himachal Pradesh , providing drinking water to them, has become one of the most polluted river in hill State making it a hazardous source of water. The river wades its way through some of the districts like Kulu, Manali, Hamirpur and Kangra. Important towns like Manali, Kulu, Mandi, Nadaun, Sujanpur are located along the river side besides hundreds of villages. The dumping of solid waste coming from various activities in these towns, open defecation by the people along the river and disposal of garbage are some of the major factors, contaminating the river water. The towns along the river further add to its pollution as several crematorium are located right on the bank of the river or its tributaries. The slate mines situated in Ravi catchments contribute 200 tonnes of solid wastes daily. The river also receives two crore litres of liquid and 650 tonnes of solid waste from 7 main towns situated in the catchment areas.

Pollution of the river Chenab near its source *i.e.* Lahaul (Himachal Pradesh) is negligible. However, its water gets very much polluted after entering into the river a number of sewerage *Nallaha,* while it moves from Lahaul to Akhnoor town via Padar, Kishtwar, Doda, Ramban and Reasi. Similarly, the river Jhelum is badly polluted with domestic sewage and waste water when it flows into Sringar city from Cheshma Verinag via Khanbal a town of Anantnag district. In fact, the Jhelum has become the dumping ground to all the liquid and solid waste of the people of the valley. The entire sewage of Kashmir valley enters the Jhelum and in the city of Srinagar the river is stinking awfully.

The Jammu city now known as old city has unique geographical situation from drainage point of view. Every drop of waste water including that of rains drains off quickly and finally reaches the river Tawi. The people of Jammu are now using bucket type of latrines and hand flushes, the waste water of which discharges directly into the drains and finally to the river Tawi. This river is, thus getting polluted more and more, and has reached an alarming stage.

Ocean Pollution

Widespread pollution has already been observed in the Indian ocean including Bay of Bengal and Arabian Sea which could affect monsoon in the Indian sub-continent. In fact, a dense brownish haze of microscopic pollutant particles suspended in air (aerosols) in Indian ocean has been found by the International Team of Researchers. These aerosoles are responsible for affecting monsoon parameters.

Aerosols primarily contain soot, sulphates, organic compounds, flyash and mineral dust. All these substances scatter incoming sunlight and reduce the amount of heat energy absorbed by the ocean by about 10 per cent. Decreased amount of sunlight going into the ocean might affect the amount of moisture evaporating from the surface of the ocean which would in turn affect the rainfall. This would also affect marine plants which depend upon photosynthesis including planktons which have a key link in the marine food chain.

Sediment load brought about by various rivers is another important factor which is responsible in causing ocean pollution. In India, more than 20 million hectares of land are affected annually by flooding due to climatic condition. The Ganga and the Brahmaputra river system carry annual suspended sediment load of about 3000 million tonnes to the sea causing considerable damage not only in the form of surface fertile soils (Singh *et al.*,2000) but also heavy metal pollution. Sediments also carry substantial quantity of nitrogen and phosphorus *vis-a-vis* pesticides, Accumulation of these nutrients stimulate unwanted growth of many organisms causing reduction in dissolved oxygen, which upsets the aquatic life. Pesticides can cause fish and animal death and some can be hazardous to human beings if they move through the food chain.

Water Pollution in Lakes

Nearly a quarter of Sweden's 90,000 lakes are acidified to some extent, 4,000 of them so severely that no fish are survived in them. In the southern half of Norway, four-fifths of the lakes are technically dead or on the critical list. Thousands of lakes in eastern United States including atleast 10 per cent of all those in the mountains, are too acidic to support fish. More than 300 lakes in Canada's Ontario province have pH value lower than 5.0.

Both saline and fresh water lakes are present in India. Saline lakes abound in arid or semi arid regions, whereas those which are extremely poor in electrolytes are abundant in humid, or temperate regions, nurtured by monsoon or winter rains. These lakes are situated at altitudes from 600 to 5600 m above mean sea level, and are experienced to climatic condition varying from cold desert of Ladakh to wet humid of Manipur (Bahadur,1999). The inflow of high silt load from glaciers, streams, rivers and catchments renders the lake water turbid and unfit for biological activity and gradually filling the lakes. The other impact is from pollution of agricultural, industrial, human and cattle wastes.

Large lakes such as the Wular and the Dal in Kashmir, the Mansar and Surinansar in Jammu, the Tsokar the Pongong Tso in Ladakh, the Rewalsar in Himachal Pradesh, Sultanpur lake in Haryana, the Nainital lake in Uttaranchal Pradesh, the Chilka in Orissa, the Mirik and Jore-Pokhari Lakes in Darjeeling , and the Kheehipiri lake in eastern Himalaya , are highly polluted.

The Wular lake once claimed as the purest, the largest natural water lake of Kashmir in Asia, is subjected to deterioration owing to nutrient flow and pollution from its catchments. The lake has been shrinking in all the three dimensions and degrading due to:

(i) with the increase in population, there is heavy rush on the land for cultivation by the inhabitants;
(ii) the entire catchment of Kashmir valley upto Sopore drains ultimately into the Wular lake, resulting into the reduction in the size and depth of the lake; and
(iii) with increasing demand for fuel and fodder vast lake areas have brought under afforestation, and such plantation while deading the flow, causes the lake water to shed its sediment load within the lakes otherwise would have been washed off through outflow channel.

The Dal lake is filled with filth and dirt by house boats and houses situated nearby. The deposition of silt by erosion and plenty of growth of weeds have greatly disturbed the ecosystem of the lake. More than about 8,000 tonnes of silt is into the lake via Telbal Nalha, converting about 0.5 km^2 area of highly polluted due to accumulation of high amount of phosphate and nitrate causing eutrophication. The water of the lake is changing colour. During August 1991 red hue was observed due to the presence of red algal mass. It was identified as chlorophyceae, which may turn the lake into a marsh. There is also a heavy metal pollution.

Like other lakes, the Mansar and Suriansar lakes have also very much deteriorated under the biotic influence of man. Cremation ground, tourists rest houses and temples situated on the banks of the lakes add much polltants into the lake. The greater heavy metal content in Mansar lake like copper, manganese, lead, cadmium and chromium has also been reported. (Gupta and Dogra, 1999). Lot of argillaceous material comes into the lakes during rainy season from the catchments, thereby, upsetting the flora and fauna of the lakes.

The major source of pollution in the Nainital lake is human faeces which are carried by leaking sewer lines or by direct pollution along the shore. The pollution of the lake is further increased by increasing number of paddle boats, from where both food and nonfood degradable wastes are thrown during boating. The lake is the main source of drinking water for the Nainital inhabitants and recently large number of Jaundice cases have been reported.

The Chilka lake, a unique gift of the God for the people of Orissa and covering an area of about 350 km^2 is turning into a lagoon or marsh due to over exploitation of its natural resources and heavy siltation. A survey conducted by limnologists has shown that the lake is losing an area of about 15 km^2 annually.

Now-a-days Ravalsar lake is passing through the most crucial period since the time of its creation. People for their selfish means have played havoc with the ecology environment of the area and made all its catchments deforested. This has resulted in the siltation of the lake. Due to eutrophication, the fishes and other fauna have been disturbed to a large extent.

Sources or Causes of Water Pollution

Like air pollutants, water pollutants arise from numerous anthropogenic and natural sources, Likewise water pollutants produced in one country may flow into others through rivers, seas and oceans, creating complex international control problems hard to solve. The important sources or causes of water pollution are as follows :

Domestic Sewage and Waste Water

The Jammu city, has a unique geographical situation from drainage point of view. Every drop of waste water including that of rains, drains off quickly and finally reaches the river Tawi. However, before 1947 when population of the city was about 50,000 the quantum of domestic sewage and waste water was much less. But with the population explosion which is presently about 10 lakhs, the amount of wastes water has tremendously increased. Moreover, the people of Jammu are now using bucket type of latrines and hand flushes, the waste water of which discharges directly into the drains and eventually to the river Tawi. This river is, thus, getting polluted more and more and has reached an alarming stage (Gupta, 1989).

The Chenab river water is practically pure at its source of origin in Lahaul (Himachal Pradesh). But by the time it reaches

Akhnoor, gets polluted sufficiently through sewerage discharges into it en route from many town. Similarly, water of the river Jhelum is badly polluted with domestic sewage and waste water especially when it flows into Srinagar city. Population of Srinagar city is progressively increasing. There is gradual encroachment in the flood plains on the river banks resulting in increase of river water pollution. In fact, the Jhelum has become the dumping ground for all the liquid and solid wastes from towns and cities.

Various other water bodies like streams and lakes both in Kashmir and Jammu are highly polluted due to this and other reasons. A great mass of pollutants in the form of domestic sewage and waste water of Udhampur town has been polluting the fresh water of the Devak stream, which is considered to be the holiest stream. Likewise, the water of Mandi and Loren Nallahs in Poonch district, is being polluted due to sewerage of the habitants residing very near to their banks. Once claimed as the purest, the largest natural water lake of Kashmir in Asia the Wular is now subjected to deterioration owing to nutrient flow and pollution from its catchments.

Similarly the water of Dal Lake in Kashmir valley was once so clear that mountain ridges were reflected in its water as in a mirror, but now the crystal clear water is very much polluted due to a number of pollutants. The garbage and effluents from the adjoining areas contaminate the water and increase the intensity of the water pollution (Gupta, 2002).

There are about 850 registered house boats and an equal number of unregistered service doongas, resident doongas and more than 1000 shikaras in the Dal Lake. These house boats provide about 3600 bed accommodation and every house boat has 2 to 3 toilets. Their flow accompanied with the sewerage from adjoining hotels/restaurants as well as from Rainawari area of the city and other nearby regions contaminate lake's fresh water. A leprosy hospital near Nagin lake is letting out its refuse directly into the lake.

The picturesque Kulu-Manali valley in Himachal Pradesh is facing threat of water pollution in the wake of widespread and haphazard tourism induced urbanisation. Sewage and debris emanating from 600 odd hotels (the number was only 10 in 1985-86) intermingle and flow into the river Beas in absence of any traditional system of sewerage disposal. The number of tourists arriving in the valley had arisen from 11,002 to 67,132 between 1975 and 1991. This had polluted the water of the river Beas as there is no proper planning for waste disposal with garbage lying everywhere along the river bank. The widespread use of polythene bags by the tourists is another environmental problem as most of these bags ultimately find their way into the river. Another major source of pollution in the Beas river is the dumping of solid waste into the river coming from a various activities into the river in Kulu, Manali and Mandi. None of these towns has proper treatment facility like incinerators.

Night Soil and Animal Excreta

The rivers, wells and other water bodies are polluted with our own excreta besides that of animals and birds.

What to talk of villagers, many inhabitants of Jammu and Srinagar cities as well as of big towns like Poonch, Rajouri, Udhampur, Akhnoor, Basholi and Kathua and others, defecate on the banks of the rivers daily in the morning which contaminates the rivers. The bacterial population of these rivers, causing water borne diseases, increases manifold as compared to their place of origin. Open defecation by people of a number of towns and village of H.P. is a common place along the river Beas.

The huge mass of deadly pollutants is daily spreading its tentacles unchecked, as vehicle after vehicle, belonging to Jammu Municipal committee, empty their stinking mass of city refuse into a huge dump. This has become a horror of helpless residents in the fifty villages living around the arc shaped land mass criss-crossed by Tawi river are about 3-4 km down stream from Jammu city. Swarms of flies arise from these dumps of refuse and fly in different directions of the villages carrying disease larvae.

The sweepers of Akhnoor town throw nightsoil in the river Chenab. This is also the practice of sweepers in other towns like Kathua and Basholi, who throw the nightsoil in the river Ravi. Open defecation by the people of several towns Mandi, Kulu, Sujanpur, Nadaun and Dehra in H.P. is a common place along the river Beas. Human faeces is also the major source of pollution in Nainital lake which are either carried by leaking sewer lines or by direct pollution along the shore. The pollution is further aggravated by the increasing number of paddle boats from which both food and non-degradable wastes are thrown into lake during boating.

Excreta of the birds, wild and pet animals like cows, buffaloes, camels and others, causes foul small in the water of the rivers and lakes. It has been estimated that waste from about 10,000 people and 1,000 horses and mules and 200 shops from Adhkwari alone either percolates or directly flows down to the Banganga river in Katra daily. The rotting of horse dung and other wastes not only create offensive odour in the water but pollute it badly biologically and chemically. According to rough estimates about 25 tonnes of horse dung enters the Nainital lake in the hills of Kumaon (Uttranchal) daily.

Industrial Wastes

The pollution of various rivers, streams and lakes by industrial wastes has been a problem since the rise of industrial revolution. The wastes and effluents from industries are generally discharged into the nearby rivers and streams to get rid of them without taking into consideration their deleterious effects.

The mushroom growth of industries in Satwari, Digiana Gangyal belt and Bari Brahamana has already caused great damage to the scenic and aesthetic view of Jammu at its entry. It has created nuisance for the public health as the industrial wastes (solid, liquid) lie around and during rains pollute the water bodies after washing away the industrial wastes.

The obnoxious smell of the distilleries at Gangyal, Samba and Bohri (Talab Tillo) and turpentine oil factories at Miran Sahib and Samba not only pollutes the surroundings but also destroys the nearby cultivable land and chokes the drains with its injurious waste water. The Desi bhattis on the Tawi islands, suburban and rural area are no less harmful. There is no proper outlet and disposal of the industrial waste even in the recognised industrial areas and the newly coming Birpur complex is another addition to the unplanned development works done by the engineers/bureaucrats. The industries are no less criminals in adding to the pollution, even the drinking water meant for labour is contaminated within the factory premises.

Though in the factories/industries at Bari Brahamana/ Gangyal and Birpur, the pollutant emission is much less yet it is in the form of refusal water having adverse effects on both human life, animals and vegetation as well as soil. The flowing of refusal water from adjoining factories of Bari Brahamana is a great threat to soil resources. It may change the soil reaction depending upon its pH.

Besides acids and alkalies, industrial wastes and discharge also contain toxic heavy metals like copper, lead, zinc, mercury, cadimium, arsenic, chromium, selenium etc. Lead, copper and zinc are drained from the rocks bearing these metals and also released from chemical industries. A number of stone-quarrying and clay supplying industries provides limestone, Permian limestone, gypsum, building stones and porcelain clay, are going on in Kishtwar area as well as on the National Highway from Jammu to Srinagar, These industries generate large quantities of solid wastes in big heaps. These wastes not only pollute the water by exudation of heavy metals having higher concentration but also the rapid generation of suspended solids in water when the heaps are washed away by heavy rainfall or flood.

Another big pollutant of water is the dumping of crushed rocks and rubbles from the various *Khads* after extraction of limestone. Pouni Khad passing through Pouni town (near Reasi) is one of the worth mentioning example in this regard A continuous extraction of limestone from this *Khad*, has choked its water which passes near the crushing establishments of the limestone.

Another pollutant of streams and river waters is the coal mines. When extraction of coal mine is done by machines, a huge mass of waste is generated and dumped outside the mine. When the rains come, it gets carried into the river. The other pollutant is coal dust which contains sufficient quantity of suspended solids besides toxic substances. This phenomenon is prevalent in Kalakote and Handwara areas of the State.

Bathing and Washing

Many of the rural as well as urban folks bathe and wash the clothes daily on the banks of rivers, ponds and other water bodies. As a result, lot of dirt and soap is released in the water. Not only this, the water is also contaminated with large number of deadly organisms producing diseases like tuberculosis, typhoid, dysentery etc., when some of the rural and urban folks suffered with these contagious diseases take bath and wash their clothes in the river water. Bathing *ghats* in main towns/villages along the river Beas (H.P) also contribute to all this pollution.

Bathing of cattle in streams, rivers, ponds is also a common practice by many of the people of the villages, towns and even in the cities. Due to this, the water is polluted with the dung present on the bodies of the cattle. More than a hundred bovine animals take bath daily in the river Jhajar Kotli, a picnic spot for school children and others. The children even drink water from there and are at risk of having all the contagious and animals borne diseases.

The shawl cleaning and carpet washing is done in Dal Lake with the belief that there are some such ingredients in the water which give extraordinary brightness to the colours and improvement in their texture. The dirt, rags and wastes produced by wholesale washing or cleaning of carpets and shawls pollutes the water to a great extent.

Crematoria

Throwing of unburnt or partly burnt dead bodies and animals carcasses into the river is a cause of pollution in our streams. Many Hindus of Udhampur throw ashes after cremating the dead bodies into the Devak stream and other either in River Chenab or in the Tawi.

Siltation

Silt is one of the components of soil with particle size 0.002 mm diameter. Silty loam and silty clay loam soils contain silt as predominant component followed by loam or sandy loam soils. The silt and sand particles are carried away from land by run-off of water during soil erosion and cause very serious pollution of water resources.

The deposition of silt can choke irrigation channels, streams and rivers. It can also affect the life span of hydroelectric projects and life of the aquatic biota, for example Salmon (edible fish) are unable to spawn on silted gravel beds.

The Jhelum flowing in Kashmir valley, is getting silted up due to soil erosion from its catchment particularly the low hills. As per the soil conservation report (Kawosa, 1989) about 80 per cent of the land area of the State is affected by soil erosion. This process is aggravating through the development of hill roads, highways, urbanisation, quarrying and mining and other development projects. The Poru catchment alone contributes 24 lakhs acre feet of silt to the river Jhelum. With slight rains, the Jhelum is in spate and brings misery to the lakhs of people of the valley. Denudation of hills/ mountains due to wholesale deforestation has increased silt flow considerably into the Wular Lake where the river Jhelum ultimately enters after receiving various tributaries. This has reduced the size and depth of the lake.

The frequent occurrence of avalanches and glaciers in Padar and Kishtwar, produce a lot of quantity of scree (fragmental and sorted material formed due to frost shattering action of sedimentary rocks and its movement under the influence of gravity) which on falling in the Chenab river makes it polluted. Moreover, due to the disturbances in ecological balance of various catchments of this river, even slight rains causes huge quantity of silt and boulders which get deposited at various places below Akhnoor bridge after the river Chenab enters the plains. The flash floods in this river on 25th and 26th September, 1988 carrying huge bed material got deposited at various places causing large scale erosion of the banks and over topping of the embankments already constructed. Similar was the case during 10th and 11th September, 1992. The river Tawi and Ujh also suffer badly due to siltation. The flood discharge of 3.5 lakh cusecs experienced in the river Tawi on 25-26th September, 1988 brought so much silt, rubbles, stones and boulders which caused large scale devastation to the flood control works. A substantial area of Jammu was inundated on both the banks of Tawi causing unfold misery to the people.

Besides three parameters viz. physical, chemical and biological causing water pollution in the Dal Lake, denudation of catchment area particularly in Telbal catchment has resulted in a significant contribution of silt. This has converted the northern part of Dal Lake into a sizeable marshyland. A lot of fine grained clays, argilaceous materials as well as sand and silt enter the twin lakes of Mansar and Suriansar from nearby Siwalik hills.

Marginally over 1935 m above mean sea level, the 1.4 km long crescent shaped Naini Lake in Kumaon Himalaya of Uttranchal has a maximum depth of 26 m. The 21 km^2 basin of the lake is surrounded by seven hills with an average slope of 19^0. The lake has, however, been plagued by seeping sewer lines along with organic and toxic wastes for the past 10 years. Besides, it is getting shallower with an average accumulation of 15,500 cubic metres of sediments.

Agriculture

Agriculture has been a victim of water pollution in many instances, but sometimes, it is also responsible for polluting water. Water pollution caused by agriculture is mainly an outcome of the following:

Cultivation

Modern agricultural practices require the use of large amounts of fertilizers and soil additives. Fertilizers applied to crops are not always fully consumed by the plants, a part of them remains in the soil being absorbed by its colloids (clay and humus). However, on dissolution with water through rainfall, irrigation and drainage, some of it is washed into the rivers and streams where they seriously disturb the ecosystem. As chemical fertilizers mainly consist of relatively simple compounds of nitrogen, phosphorus and potassium (major primary plant nutrients), their run-off coefficients vary depending upon the solubility of the fertilizer itself, the rate of adsorption by the soil and the rate of decomposition. It is generally believed that the run-off coefficient of nitrogen is 5-30 per cent and that of phosphorus is 0.5-5 per cent. The former having a high run-off coefficient tends to cause the eutrophication of lakes, ponds and other water bodies. The Dal Lake has shown many signs of eutrophication (Zutshi & Khan, 1988) as a result of accumulation of nitrates in the water. Drinking of such water by cattle or human beings leads to a serious disease methaemoglobinemia which can cause severe damage to respiratory and vascular systems.

Phosphatic fertilizers are fairly persistent pollutants and when discharged into water they support luxuriant growth of algae. Excessive growth of algae cause depletion of dissolved oxygen.

The phosphorus concentration of inshore sites of Dal Lake is quite high because of the discharge of domestic waste which contains nitrogen and phosphorus load to the tune of 359 and about 40 tonnes year^{-1}.

The cultivation done on the floating lands of Dal Lake is a vital factor that threatens this water body. The methods of cultivation are very crude adding filth and dirt to the lake.

Agricultural Chemicals

No doubt man has brought about a Green Revolution in agriculture with the help of high yielding varieties of various crops responding to higher doses of fertilizers. But at the same, these are prone to more attack of diseases and insects and, as such, a variety of chemicals like insecticides, fungicides and bactericides are used to save the plant from the assaults of the pests. These chemicals are considered as a source of peculiar pollution which cannot be overlooked.

The characteristics of agricultural chemicals as water polluting substances lie in that :

(i) Almost all of them are special compounds not occurring at all in nature and their behaviours/changes when they are discharged into the natural environment.

(ii) Many of them influences the human body because of their inherent properties.

(iii) When they are ingested into the human body, they tend to accumulate in it.

(iv) Some of them are very staple and remain in soil, human or animal body for a long time and thereby, afflicting them considerably.

Due to cultivation of crops and development of orchards around the Dal and Wular lakes, the use of insecticides and chemical fertilizers has increased which in turn get washed down into the lakes. Similarly agricultural chemicals can drain off into the rivers. They then can kill the phyto- and zoo- plankton. These chemicals can enter the human system through food or contact and become health hazards. Some of these chemicals can also indirectly come into the human system through animals which inject the plants.

The widespread scab disease prevalent in the apple orchards of Kashmir and temperate regions of Jammu requires as many as eight dustings/spraying of insecticides in one season, and causes alarming water pollution. On an average, their amount works out to be 6,000 to 12,000 gha^{-1} of pesticides (Kawosa, 1989). The extensive and continued use of these chemicals have a deleterious effect on the water resources of the State and the soil and the atmosphere. This is also true for Himachal Himalayas. With increase in apple farming in Kulu-Manali valley and Kothkhai and Shalli area in Shimla and Sirmour respectively, there is more use of chemical fertilizers, weedicides, fungicides and insecticides. This has caused a potential threat to the soil environment as well as alarming water pollution.

Weeds

Of the serious problems of the Dal and Wular lakes, is the rapid growth of aquatic and terrestrial weeds along their peripheries. The weeds have not only deprived their recreational and aesthetic splendour but has seriously upset biological balance of lakes' ecosystems. The aesthetic value of Mansar and Suriansar lake, of Jammu is seriously threatened by the growth of undesirable weeds and shrubs.

Hazards of Water Pollution

As discussed, both inorganic and organic pollutants are present in the soluble, insoluble and colloidal forms in the water. Their corresponding effects are briefly mentioned below:

Suspended Impurities

Suspended impurities in the form of bacteria and viruses which cause diseases. Algae, protozoa and fungi are the other organisms which remain in water suspension and cause odour, turbidity, colour and even clogging of drinking water pipes. Physico-chemical limits for drinking water are presented in Table 3.1.

The people living down streams of the river Tawi are totally dependent upon river water for drinking, washing and bathing, have suffered from a number of diseases. A number of families living in the village Barjala near Niki Tawi where the Municipal Committee dumps the refuse, is one of the example how the pollutants are running the health of the villagers. A majority of the houses in the village had at least one patient either suffering from diarrhoea, gastritis or high level fever. Similarly, the other main rivers of Jammu and Kashmir viz. the Chenab, the Jhelum, the Indus and Ravi are also deteriorating fast in the quality of their water.

In big cities like Jammu and Srinagar themselves the swamps and stagnant water in many places especially during rainy seasons have become the breeding places for mosquitoes and flies causing diseases such as typhoid, cholera, malaria and gastroentritis.

While going to Vaishno Devi Yatra every body likes to take a holy dip in the river Banganga as it is believed to give relief from heat and relaxes tired body. But this water is not fit for drinking or even bathing. Very high rate of skin diseases has been found besides gastrointestinal. The water borne diseases include viral hepatitis, paratyphoid, dysentery, amoebiasis, round worms, thread worms, trachoma and bronchial infections etc.

Since the water of Nanital Lake is the main source of drinking water for the inhabitants of Nainital so large number of Jaundice cases have been reported. Besides, more than 90 per cent of Nainital population suffers from stomach ailments like amoebiasis.

The red algal moss has been found to have appeared during August 1991 in the Dal Lake. This has been identified as chlorophyceae, cyanophyceae and bacillariophyceae and is responsible to impart odour, colour, smell of the water. This is, of course, due to action and interaction of physical and chemical properties of the polluted water of the lake.

Dissolved Impurities

Dissolved impurities include various salts and metals. Carbonates and bicarbonates of calcium and sodium produce alkalinity and temporary hardness of water, whereas chlorides and sulphates of magnesium and other produce permanent hardness of water. Iron in its oxide form gives bad taste and red colouration. On the other hand, manganese imparts black or brown colour in the water, helps organic growth, while lead acts as cumulative poison.

Table 3.1: Physico-Chemical Limit for Drinking Water

Parameter	Acceptable	Cause for rejection
Turbidity	2.5	10
Taste and odour	Objectionable	Objectionable
Total dissolved solids mg L^{-1}	500	1500
Chloride mg L^{-1}	200	1000
Fluoride mg L^{-1}	1.00	1.5
Calcium mg L^{-1}	75	200
Iron mg L^{-1}	0.1	1.0
Copper mg L^{-1}	0.05	1.5
Phenol mg L^{-1}	0.001	0.002
Arsenic mg L^{-1}	0.05	0.05
Chromium mg L^{-1}	0.05	0.05
Lead mg L^{-1}	0.1	0.1
Mercury mg L^{-1}	0.001	0.001
Colour (Platinum cobalt scale)	5	25
pH	7.0-8.5	6.5-9.2
Total hardness $CaCo_3$mg L^{-1}	200	600
Sulphate mg L^{-1}	200	400
Nitrate mg L^{-1}	45	45
Magnesium mg L^{-1}	30	150
Manganese mg L^{-1}	0.05	0.5
Zinc mg L^{-1}	5	15
Cadmium mg L^{-1}	0.01	0.01
Cyanides mg L^{-1}	0.05	0.05
Selenium mg L^{-1}	0.01	0.01

WHO 1984, *Guidelines for drinking water quality* Vol-1, Recommendation World Health Organisation, Geneva, pp. 1-130.

Mercury is a serious food contaminant and affects human nervous system. Lead affects enzymes and impairs cell metabolism. Excess fluoride in drinking water can lead to fluorosis, a disease that surfaces in a variety of symptoms and can cripple a person. Severe and widespread gastro-intestinal problem viz. loss of appetite, stomach pain, intermittent diarrhoea, chronic constipation, gas formation and bloated feeling in the stomach caused due to drinking fluoriated water have been confirmed. White, yellow, brown and black tooth enamel discolouration either in spots or in horizontal streaks is the sign of occurrence of dental fluorosis.

In Jammu and Kashmir, endemic fluorosis has been observed in two blocks viz. Ghata and Bhagua of Doda district. In Ghatta, 90 per cent people are affected with dental fluorosis, 60 per cent with gastro-intestinal and 15-20 per cent with skeletal fluorosis while in Malwas and Mokhli villages 100 per cent people are affected with dental fluorosis.

The chemical analysis of ten water samples from Loren and Mandi (Poonch district) *Nallahs* indicated that the values of calcium and magnesium varied from 9 to 47 ppm and 0.005 to 0.04 ppm respectively with averages of 26 ppm and 0.019 ppm (Fotedar, 1993). The permissible limits of these elements for drinking water supply are 75 and 30 ppm (WHO, 1981). Accordingly, the water in Loren and Mandi is considered nontoxic in respect of calcium and magnesium. However, the concentration of iron, is higher (0.1 ppm) but the manganese is in the safe limit. In another study carried out by Fotedar and Singh (1998) on 30 samples of Tawi water in respect of 10 trace and four major elements revealed that their concentration is non-toxic, exception being vanadium which has been found slightly higher in concentration than the permissible limits.

Heavy metal pollution of different inland lakes of northern Himalayas has been studied (Goverdhan, 1989). Generally, the heavy metals concentration of different lakes was dependent on the dumping of garbage, sewage and sludge, municipal waste and industrial effluents and in coming and outgoing sediment water and nature of geological formation. The amount of heavy metals has been shown in Tables 3.2 to 3.4. The chemical analysis of water samples from Khol and Dedani *Nallahs* in Doda district (Jammu & Kashmir) in respect of cations and anions reveals that all the elements present in water of both the *Nallahs* are below permissible limits (WHO, 1971) except Pb which is present in more concentration than the premissible limits. However, with usual hygienic treatment the water of these *nallahs* can be considered safe for human consumption.

Table 3.2 : Heavy metal concentration in the clay fraction of various lake sediments of Northern Himalaya in ppm

Lake	Cu	Mn	Cd	Zn	Co	Pb	Ni	Cr	Fe,%	PO_4,%
Dal										
Boddal basin	142	1148	0.96	184	84	56	86	94	0.56	1.04
Gagribal basin	103	1085	0.87	146	76	48	77	86	0.42	1.64
Hazratbal basin	89	1123	0.88	180	73	39	69	85	0.45	1.53
Nagin basin	94	945	1.05	165	68	42	82	88	0.49	1.38
Anchar	64	875	0.74	103	41	54	68	76	0.35	0.96
Wular	89	698	0.36	159	32	58	72	84	0.32	1.08
Khanapur	104	748	0.42	178	30	62	73	68	0.34	0.86
Trigam	140	1085	0.38	164	23	68	75	64	0.36	0.64
Tilwan	105	878	0.54	138	27	50	69	65	0.31	0.82
Nilnag	45	578	0.62	142	26	43	72	63	0.35	0.78
Manasbal	82	979	0.48	139	24	49	65	72	0.36	0.82
Saunisar	108	642	0.88	194	38	49	84	92	0.46	0.82
Mansar	114	954	0.74	132	22	36	73	84	0.32	0.68
Alipathar	64	472	0.56	149	31	39	69	86	0.36	0.74
Jarsar	34	398	0.48	132	33	48	72	87	0.34	0.67
Naini	175	1258	0.92	163	42	32	92	96	0.58	0.83
Marsar	58	678	0.45	103	36	43	68	84	0.42	0.74
Kounsernag	69	814	0.63	145	35	45	83	88	0.32	0.58
Sheeshnag	34	413	0.39	189	33	31	88	89	0.33	0.43
Gangabal	39	345	0.59	114	34	28	79	88	0.34	0.52
Vishansar	45	438	0.48	145	38	36	74	83	0.36	0.54

Source: Goverdhan, V. 1989., *Indian J. Environ. Protec.*, 9(7) : 531—536.

Table 3.3 : Heavy metal concentration in sewage sludges of different basins of Dal Lake complex, in ppm

Dal lake basin	No. of sample tested	Cd	Zn	Cu	Pb	Ni	Cr
Boddal	16	32	1240	685	384	86	85
Gagribal	16	43	1185	714	412	80	104
Hazratbal	16	48	1494	854	425	108	112
Nagin	16	21	1248	654	210	68	96

Source : Goverdhan, V., 1989. *Indian J. Environ. Protec.*, 9(7) : 531–536.

The concentration of heavy metals like cadmium, lead and chromium has been found to increase when compared with the data reported earlier (Goverdhan.1989).Their contents were 0.03, 10 and 31 ppm in the water of Suriansar lake as compared with the data record these days, the values of which are 0.04, 13 and 37 ppm (Gupta, 2002).

Control Measures

Water pollution, a major cause of loss of human, animal and plant life can be prevented by the following methods :

(i) The catchment areas of various rivers and lakes must be fenced in order to restrain the unabated overgrazing of the ground cover and re-afforestation programmes should be taken up in order to check the siltation in rivers and lakes. This will ultimately help in reducing the eutrophication of the lakes.

(ii) Composting involves the biological stabilization of solid matter either under aerobic or anaerobic conditions. The end product of composting is an organic matter which has beneficial value as a soil manure. Selected nutrients like phosphorus or nitrogen can be added to the compost for enriching the manure in these nutrients as well as to accelerate the process of decomposition during composting.

(iii) As far as possible industries should be away from cities and safe disposal of toxic substances must be ensured. Efforts

Table 3.4 : Heavy metal concentration in the water of different lakes of Northern Himalaya, in ppm

Lake	Cu	Mn	Cd	Zn	Co	Pb	Ni	Cr	Fe,%
Dal									
Boddal basin	34	245	0.29	64	23	42	36	36	0.26
Gagribal basin	36	286	0.26	49	14	38	48	42	0.24
Hazratbal basin	41	296	0.14	67	18	28	47	36	0.22
Nagin basin	32	265	0.23	63	12	24	45	38	0.28
Anchar	24	194	0.08	53	16	8	32	24	0.14
Wular	16	185	0.14	47	11	9	30	21	0.17
Khanapur	19	176	0.12	46	12	10	28	24	0.18
Trigam	24	194	0.06	39	13	12	27	26	0.22
Tilwan	33	243	0.08	46	14	13	26	28	0.16
Nilnag	23	232	0.07	41	15	16	32	29	0.16
Manasbal	24	249	0.10	48	13	18	34	30	0.17
Saunisar	19	132	0.03	32	14	10	24	31	0.24
Mansar	13	136	0.02	28	13	9	26	28	0.23
Alipathar	12	106	0.06	24	18	8	30	32	0.18
Tarsar	18	114	0.08	23	16	11	24	28	0.18
Naini	44	286	0.24	42	43	34	22	46	0.10
Marsar	26	106	0.06	19	15	12	21	32	0.12
Kounsernag	13	104	0.08	18	17	11	23	33	0.13
Sheeshnag	12	106	0.09	16	18	13	28	26	0.08
Gangabal	13	109	0.07	17	12	12	26	27	0.04
Vishansar	15	144	0.06	14	14	14	28	28	0.20

Source : Goverdhan, V. (1989), *Indian J. Environmental Protection.*, 9(7) 531–536.

should be directed towards minimising chemical contamination and for ensuring safety from accidents.

(iv) As a precautionary measure, drinking water should be boiled. Chlorination has many side effects and should be avoided.

(v) Ultraviolet irradiation consists of passing ultraviolet rays from mercury vapour arc lamps. The astonishing results of this method is shown as under :

Pathogens	***Killing time (Sec)***
Cholera	10-15
Dysentery	13-20
Tetanus	20-60
Staphylococcus	5-10

(vi) The use of agricultural chemicals *i.e.* fertilizers, insecticides, fungicides, rodenticides, weedicides etc. should be minimised or preferably avoided.

(vii) Avoid open defecation particularly in the banks of the streams, rivers, lakes and springs.

(viii) Conventional treatment methods for public water supply consist of pre-chlorination, aeration, flocculation, filtration and post-chlorination. Chlorination is health hazard and should be avoided.

Chlorination not only removes bacteria but also eliminates taste, colour and odour alongwith control of algae, fungi and protozoa, but it has many side effects.

The principle behind aerations is to mix with water to increase oxygen content and remove carbon dioxide. It is also done to remove iron from surface water. It is a safe method.

Finally divided suspended matters, both organic and inorganic which cause colour and turbidity, do not settle easily through sedimentation, the objective of coagulation is to convert these into easily settleable size by reducing Zeta potential.

Filtration aids in the removal of taste, odour, colour, turbidity, iron and manganese and to large extent of bacterial load.

(ix) Weeds of the lakes must be removed either by mechanical cutting and harvesting or manually.

Suggestions and Priorities for Future Improvement

(i) A high priority should be accorded to educate the citizens/ rural people about environmental risks, the economic and health dangers of sources of pollution. We must lay emphasis on promoting awareness about fighting pollution in schools, colleges, government and private institutions.

(ii) Another priority of the day is to follow the solid waste management aspects strictly and properly. There are four such aspects which include :

- Proper storage at or near the point of generation.
- Effective collection.
- Street cleansing.
- Right disposal.

To achieve this goal, we have to depend on the quality of management. It is necessary to deploy a complete set of technical skills which derive from several professional disciplines. These include civil and mechanical engineering, transport organisation, land-use planning and economics. Thus solid waste management is a team work and whatever the original discipline of the manager, his prime qualification should be his specialised background in the field.

(iii) In rural areas, the biological wastes of animals and homes is converted into manure and used in the land to make it more fertile. However, these wastes are stocked in the fields and the process of conversion to manure is left to be completed naturally.

It is undesirable practice as this creates water pollution besides loss of nurtrients. Hence, these wastes should be collected in pits or tanks. If these are collected and dumped in tanks, Gobar gas can be produced to be used as a source of energy. Alternatively these wastes should be covered with 20 cm to 30 cm soil.

(iv) People in the towns, streets and bylanes who throw the wastes out of their houses into the lanes, open areas and in drains, must be punished. Certain laws and rules must be enacted.

(v) The dirty and polluted water which is thrown by the administration in various streams or rivers, should be cleaned before being released in them.

(vi) Industrial estates and clusters of small scale industries should be scattered throughout State. This will ensure equilibrium in the distribution of population and equibalance in emitting the pollution gradients and its natural control measures.

The establishment of clusters of small scale industries in different rural areas will also stop the flow of pollution from the villages to the towns and cities, thus relieving them of excessive pollution.

(vii) People in the rural areas should be educated in the implementation of pollution control measures. This should be achieved promoting development and adopting of cleaner technologies, including environmentally friendly bio-technology.

(viii) Rural sanitation programme must be started on large scale. The strategy of the sanitation programme should be on integrated package of practices towards a holistic management of the liquid and solid wastes, drainage, personal hygiene environment and general sanitation in the rural areas.

It is also proposed to make this programme more broad-based by adding the components of biogas, solar energy immunisation and child nutrition through functional coordination build-up with departments like Health and Social Welfare. It is further proposed to develop a few model villages for emulation by agencies in other areas.

(ix) It is proposed that the waste water from the toilets and wash basins can be discharged into the sewer and after certain treatments can gainfully be used for irrigation. This is known as recycling of water. In fact, the nutrients present in treated waste water are an advantage for agricultural production. Similarly waste water from bath room, kitchen etc. can be subjected for minimal treatment for making it fit for flushing toilets and watering lawns.

Industrial processed water can be recycled for cooling purposes, since cooling can tolerate low quality water.

Impact Point to Remember for Water Pollution

- Water constituting about 70 per cent of earth's surface is so much polluted that today the water effects are global based.
- Almost all big rivers and lakes of the world are so much polluted that cannot be processed for drinking purpose.
- Gangotri the birth place of the river Ganga today itself suffers from a multiproned invasion in the form of erosion, degradable and non-degradable waste.
- The water of Ganga which was supposed to have magical powers has now itself become a part of the pathological cycle of bacteria and protozoa. Similar is the case with the river Yamuna.
- Water pollution of these rivers threaten the livelihood of millions who derive subsistence through fishing and boating.
- Out of 3-6 per cent of the world's fresh water, more than 75 per cent is locked away in glaciers and polar ice caps.
- Health and environment experts are urging policy makers to heed the impact of population density and fresh water scarcity on the spread of diseases.
- Presently, more than 1.2 billion people cannot drink water without risk of diseases or death.

- Storm run off of water which carries pesticides, organic wastes, irrigation water containing salts and sediments carrying nutrients, cause contamination of surface water.
- Ground water gets contaminated as a result of leaching of nitrates due to excessive use of nitrogenous fertilizers.
- The World Health Organisation has recommended to supply 150 litres water per person per day for domestic, drinking, washing and cooking requirements.
- The Government of India has suggested a norm of 180 litres per capita for all domestic needs.
- But it has not been possible to supply atleast 100 litres day^{-1} even in the urban areas.
- Food and Agriculture Organisation has estimated that the water requirement in any country becomes double in every 21 years due to population explosion both in human beings and animals.
- In India about 50 million people are affected due to excessive fluoride, arsenic, nitrate, iron, manganese, and carbonate and bicarbonate in drinking water.
- Out of 50 million people, 30 million alone spread over an area of 3 lakh km^{-2}, mostly in the Western and north-western regions of the country.
- Typhoid, chlorea, hepatitis and gestroenteritis are the main water borne diseases.
- Unsafe water is responsible for tape worms, hook worms and thread worms.
- The municipal wastes cause viral diseases and viral diarrhoea.
- High concentration of arsenic in ground water causes anaemia, skin cancer, lung diseases and many others.
- Excess of fluoride causes fluorosis (a bone disease) severe anaemia, stiff joint, painful restricted movement, kidney failure, loose muscles and molten teeth.
- The catchment areas of various rivers and lakes must be fenced to restrain the unabated overgrazing the ground water.
- Reafforestation programmes should be taken up on priority.
- Diversion of sewage channels and construction of waste

treatments should be started, composting of solid organic wastes should be encouraged.

- Over extraction of ground water should be totally banned.
- People in the rural areas should be educated in the implementation of pollution control measures.
- Avoid open defecation particularly on the banks of the streams, rivers and lakes.
- Conventional methods for water supply such as aeration, flocculation, filtration should be preferred rather than chlorination.
- Ultraviolet irradiation to the drinking water for killing pathogens should be done.
- Use of agricultural chemicals should be minimised or preferably avoided.
- Damodar Action Plan on the lines of Ganga Action Plan should be started to prevent the pollution of the Damodar river.

REFERENCES

Bahadur, J. (1999). Himalayan ecohydrology and emerging topic. In *The Himalayan Environment*. New Age International (P) Limited Publishers, New Delhi, Bangalore, Culcutta, Chennai, Gwahati, Hyderabad, Lucknow, Mumbai.

Fotedar, B.K. and Singh, V.P. (1998). "Heavy metal pollution in the river Tawi." Personal Communication, Department of Geology, Jammu University, Jammu.

Fotedar, B.K. (1993). Chemical analysis of water samples from Loren and Mandi (Poonch District). *Daily Excelsior,* Volume, 29.

Govardhan, V. (1989). Heavy metal concentration in the water of different lakes of northern Himalayas. *Indian J. Environmental Protection*, 9 (7) : 531-536.

Gupta, R.D. (2002). Water pollution in J&K. *Daily Excelsior.* 38 (201). 7

Gupta, M.L. (1989). Clean Environment and Jammu. In Souvenir (2nd National Symposium on Hydrology) J&K Engineering Department, Hydraulic wing.

Gupta, R. D. (2002). Save shrinking Suriansar Lake. *Daily Excelsior,* 38(243): 9 (*Sunday Magazine*).

Gupta, R.D. and Dogra, S.R. (1999). Save Mansar Lake from Extinction. *Daily Excelsior* 12th April (1999) issue.

Karunakaran, T. (2000). Globalisation of environmental issues. *Environment and People.* 7 (7): 15-16.

Kawosa, M.A. (1989). State of Environment in Jammu and Kashmir. *Bhagirath*, 36 (4) : 184-188.

Mohammed, N.S. (2000). The cultural basis and environment crises in Canada and India. *Environment and People* 7 (7): 23-25.

Sharma, S.; Vora, J.J.; Marfani, M.F and Joshi, J.D. (2000). Fluoride Problem in north-west India. *Environment and People* 7 (7) : 7-9.

Singh, K.K.; Patrin, M.L. and Sharma, H.C., (2000). Environmental protection and sustainable agriculture. *Intensive Agriculture* 37 (1-2) : 12-16.

Subba Rao, M.V. (2001). Water and Rain Water Harvesting Management. *Environment and People* 8 (6) : 10-12.

WHO. (1981). *Health For All.* Serial No.4.

WHO. (1971). International Standards for Drinking Water, Geneva.

Zutshi, D.P. and Khan, A.U. (1988). Eutrophic in the Dal Lake, Kashmir. *Indian J. Environment and Health.* 30 (4) : 348-354.

4
NOISE POLLUTION

Concept of Noise Pollution

Noise is one among the several pollutants and polluting agents, having potential of causing a host of diseases, ranging from mild headache to permanent hearing loss and heart attack. As the noise pollution level is reaching the alarming heights in every mega city, so it is the high time the authorities put a curb to the sources of noise pollution. However, prior to dealing with the noise pollution we must be clear in our mind as what is noise and what is pollution.

Definition of Noise

The word noise is derived from the latin word *nausea*. There are several meanings of noise which when collectively considered give us a clear idea as to what constitutes noise. Inspite of a number of purposeful meanings, it still remains a subject, that is open to interpretation. For instance, in a political rally the participants are exhilerated to hear their menacing interloping in their peace. Similarly, what the youngs may consider lovely music, may become hell to the old and sick. Hence, basically noise is a sound that pleases the ear is music and that which causes irritation and annoyance is noise. A few definitions will give an idea "Sound without value", "Sound that is not desired by the recipient", "Wrong sound at any time or place", and so on.

"Noise" as defined by the World Health Organisation (WHO), is undesirable and unwanted sound."

"Noise is generally defined as the sound that is disturbing to the normal level of human perception of sound". The degree of disturbance, however, varies widely with the sensitivity of the person *vis-a-vis* factors such as age, health of the hearing mechanism and the nature of the stress and strains undergone by the individual.

Webster defines noise as "a sound that lacks agreeable musical quality or is noticeably loud, harsh or discordant".

What Constitutes Noise Pollution ?

Though the menace of loud noise and echophony has haunted man from time immemorial; Robert Alex Baron (1964) was the first who showed that noise which forms an inseparable part of every day living for many people, has damaging physiological and psychological effects. So loud noise which is omnipresent especially in the crowded cities of the country where there is no legal or moral curb on noisy automobile engines and use of blaring horns and high pitched conversation, has become a serious threat to human environment. Although, most people are unlikely to experience high noise level, it may be annoying or disturbing to them. Hence, noise pollution may be defined as an "unwanted sound which gets dumped into the atmosphere to the annoyance of its recipients".

Measurement of Noise

Noise is measured by several complex systems but probably the best known unit of measurement is the decibel (dB). The decibel measures a sound intensity, or in its loudness to the ear. Indeed, the dB is a relative unit of gradation; to say that a sound is 60 dB means that it is 60 dB more intense than a sound standardised at a reference level, say zero.

The decibel scale is logarithmic *i.e.* 60 dB are 10 times more intense than 50 dB, 100 times more intense than 40 dB and 1000 times than 30 dB. A jet takes off at close proximity is 140-150 dB which is 10,000,000 times more intense than 50 dB of normal background noise in a neighbourhood.

Other Units for Measuring Noise

The intensity of noise can also be measured in "bels"– a unit named after Graham Bell. A dB is one tenth of a bel. The threshold of hearing is zero decibel, but it is only a theoretical phenomenon– accomplished in experimental conditions. What is called "silence", in general terms, it is normally around 25 dB. Noise inflicts pain at about 120 dB but severe pain and temporary loss of hearing are inflicted after only a few minutes at 140 or few seconds at 150 dB or more.

Loudness of sound can too be expressed in sones. One sone is equal to the loudness of 40 dB at 1000 hertz. The range of sound vibration below 16 Hz is infrasonic and that above 20,000 Hz is ultrasonic for the average person. But some persons can hear frequencies that others are not able to detect.

Sometimes sound is expressed in psychoaccoustic terms, the phon. It takes into account intensity and frequency. A sound level of 92 dB at 20 Hz has a loudness of 40 phons.

Human Hearing Intensity

The area of human hearing extends in intensity from 0 to greater than 120 dB. We hear ordinary dialogue or conversation at 50 dB but a sound level of 75 dB is considered moderately loud and 95 dB very loud. The noise of a jet registers at 11 0 dB, which is considered to be uncomfortably loud. At 140-150 dB, noise is painfully loud. On this basis, metropolitan cities like Delhi, Bombay, Kolkata and Chennai are excessively noisy. A recent study made by the institute of Road Traffic Education, indicates that Delhi is the noisiest city in India with 83 dB during day and 77 dB at night, followed by Kolkata (82 and 75 dB), Bomaby (80 and 74 dB) and Chennai (77 and 73 dB).

Sources of Noise Pollution

There are two main sources of noise pollution *viz*. Industrial sources and non-industrial sources, which are elaborated here as under:

Industrial Noise Pollution

In industries, noise is the by-product of energy which is converted from big machines used in various mills, foundries and other factories. Hence, industrial complexes heavy machinery working at high speed, produces much of din and the loudness of sound intensity is high which cannot go without passing its damages. Industrial noise, also

Industrial noise is mainly concerned with the industrial workers. Owing to past experience, industrial estates are established far away from residential areas. However, illegal cottage industries in residential areas can cause noise pollution for other residents. Actually noise poses twin challenge to industry. It affects the health of the workers and lowers productivity. Some experts even claim that the losses caused by hearing impairment among industrial workers are greater than those caused by any other occupational illness. (Table 4.1).

A recent study conducted by the staff of Post-Graduate Medical Sciences in Chennai shows that every fourth industrial worker, among those surveyed in Chennai, Cochin, Coimbatore, Madurai and Trivandrum, has induced noise hearing loss. Similarly study conducted by National Physical Laboratory, New Delhi has evinced that nearly 12 per cent of the Kolkata's residential areas have become unfit for human habitation due to noise pollution caused by industrial, cities traffic and Metro Railways.

Durgapur too another part of West Bengal is facing a serious noise pollution in its industrial belt. Specialists in ear, nose and throat diseases have asserted that number of patients afflicted with ear diseases have increased by 30 per cent in this belt. Besides, blood cancer, particularly among children is causing a great concern. About 2 per cent of the people are suffering from early sign of cancer.

Open coal mines in coal bearing areas of Raniganj, Asansol and Durgapur use diesel sets which contribute to the noise pollution. Similar is the case in Handwara and Kalakote areas of Jammu and Kashmir.

Non-industrial Noise Pollution

Non-industrial noise pollution sources mainly consist of transportation noise generated by the vehicles, railways and aircrafts, machinery and gadget noise generated by cement factories, concrete mixers, crushers and many others. Use of musical instruments and crackers in religious festivals, radios, transistors and loud speakers, televisions

Table 4.1: Noise levels from various sources

Noise sources	Noise levels (dB)*
Industrial	
Near Large Gas Regulator, as high as	150
Foundry Shake-out Floor, as high as	128
Automobile Assembly Line, as high as	125
Large Cooling Tower (600 m)**	120-130
Construction and Mining	
Bulldozer (10 m)	90-105
Oxygen Jet Drill in Quarry (20 m)	128
Rock Drill (Jumbo)	122
Transportation	
Jet Take Off (100 m)	130-140
Diesel Truck (200 m)	85-110
Passenger Car (25 m)	70-80
Subway (in car or on platform), as high as	110
Community	
Heavy Traffic, Business Area, as high as	110
Power Lawn Mower (5 m) as high as	92-98
Barking Dog (250 m) as high as	65
Normal Conversation	60
Night Time Urban Noise	50
Household	
Hi-Fi in Living Room as high as	125
Kitchen Blender	90-95
Electric Shaver	75-90
Inside Room (Day time)	40
Inside Room (Night time)	30

*dB = 0.002 μ bar.

** Figures in parentheses indicate listening distance.

Source: Ghaffari, H. 2000. *Environment and People* 7(5) : 12-18.

and cables, electric generators, are the other sources of non-industrial noise pollution. These are discussed one by one, hereunder:

Transportation Noise

Of the transportation noise vehicular noise is one that affects the urban population the most. This consists of engine and exhaust noise, indiscriminate use of horns, use of pressure horns, and auto repair workshop.

The sharply rising number of vehicles in Delhi estimated to reach 40 lakhs by 2001, which shall result in the steady increase in noise pollution levels as well. A survey conducted by the State pollution control board indicated that noise in mega cities like Bombay and Kolkata is as high as 95 dB, but of which the vehicles contribute 70 to 85 dB. The increased number of vehicles from 10,615 during 1971 to 85,000 by 1991 and now about 3 lakh has contributed much to the noise pollution in Jammu city (J&K State) especially at traffic clogged areas like Parade, Jewel Chowk and Bus stand.

Noise from railways and aircrafts are of major concern in big cities. Near railway stations, the intensity of sound is much loud and becomes a physical irritant to leave its diverse consequences after long exposures. In fact, the sound waves travel into the atmosphere and pollute the entire environment, and everyone feels uncomfortable in this environment. The noise level of train is more than 150 dB.

Aircrafts and supersonic without which we cannot think of modernity or development, continue generating huge levels of decibels round the clock. They have reached dangerous proportions as to have begun to prove a health hazard to the human beings residing at the airports. Because of constant exposure to sound level of 110 dB of aircrafts in the aerodromes of Kolkata, Delhi and Bombay, the people living near these aerodromes have developed hearing problems. Though in Jammu and Kashmir this noise pollution is not so oppressive as yet, it can be a main threat to the posterity when air traffic increases.

Noise from Machinery and Gadgets

This includes the noise which is produced by construction machinery like concrete mixers, cement factories and crushers.

Demolition and construction for urban renewal and expansions always make the urban man a victim of noise pollution. During demolition of old sites and construction of new buildings in the new established colonies in big cities, huge and heavy machines are employed which generate lot of noise. Stone crushers, cement, sand stone mixers are the glaring examples which have become a common scene not only in cities and big towns but also in the villages where construction work is in progress.

Pyrotechnics

Pyrotechnics like crackers and bombs are sources of unbearable noise generation during festivals like Diwali, Id, Baisakhi etc. and other celebrations like weddings, marriages, religious and social functions. The use of blasting material in construction of roads, civil works and quarries also causes lot of noise.

Use of Radios/Transistors and Loud Speakers

The noise created by the tape recorders and stereos which are played with impunity in markets, matadors and buses without any consideration for the travellers.

In the morning the *bhajans* over the loud speakers from nearby temples or *shabad kirtans* from Gurudwaras are the other sources of creating noise. The sounds from both the sources keep in criss-crossing one's ears and nothing is heard clearly. One is simply subjected to an irritating noise which makes one's morning rather uncomfortable. The problem is all the more acute for the students who get up in the morning to memorise their lessons.

Evenings are equally noisy and one can listen to loud shrills of some Moulvi from nearby mosque who in a bid to offer prayers, shouts hoarse over the loud speakers. As a matter of fact, offering prayers over the mike from the religious places has become the order of the day and these go unchecked irrespective of legal constraints linked with the use of loud speakers. Prayers are not offered but simply blared over loud speakers to the discomfort of residents around.

Truely speaking, the use of loud speakers has become virtually traditional so much so that no social function whether a marriage or religious ceremony is considered complete without echophony. The places of worship indeed have been changed into noise generators instead of dispensers of peace and tranquillity.

Use of TV and Cable

The TV sets with their loud intensities produce lot of noise pollution. As big towns/cities as well as mega cities have now been dominated by cable TVs. From cables and network from Doordarshan, the following TV sound intensities chart might be of some help to them. It is:

1. 70 dB for TV sound at medium volume.
2. 80 dB for impulse noise of advertisements at the same volume.
3. 86 dB for the film songs at same volume.
4. 82 dB for fighting scene sounds.
5. 54 dB for minimum TV volume.
6. 94 dB for maximum TV volume.

Although aforesaid data were reported by Magna Publications, formulated after a study conducted by Bombay Municipal Corporation, it might obviously hold true for TV sounds at all places.

Use of Electric Generators

Because of irregular supply of electricity in towns and cities, the use of electric generators has become very common not only in social functions, marriage/betrothal ceremonies but also in shopping complexes and other organisations. Whenever the power goes out, the shopkeepers use electric generators as substitutes thereby causing tremendous noise and badly harm the pedestrians.

Use of Musical Instruments in Festivals

In various places of India, a number of sound producing instruments are used during celebrations of festivals, religious, cultural and social activities. The Bhangra of Jammu during Baisakhi fair, the Kud dance of Bhaderwah (Jammu) and Losar dance of Ladakh, are always done with beating of drums. The folk dances of Nagaland, Kulu and Chamba (Himachal Pradesh), Gidda dance of Punjab are not only conducted with beating of drums but a number of other musical instruments like conch-shells and "brass surma" are also blown during these functions. The noise so produced is tremendous which pollutes entire environment.

In some civilised regions also where people like the Western music or pop music, noise is certain. However, the amount of noise produced, depends upon the singer.

In the villages of north Himalayan States (Jammu & Kashmir) Himachal Pradesh and hill terrain of Uttranchal *vis-a-vis* in those of Punjab, the wrestling is very popular game and it is conducted at the beat of drums. In certain cases, the drums are beaten very loudly to mark the success of wrestler. Same is true for the rural games such as bull fight, cock fight, horse races etc. During the kite flying competition the drum beaters are paid according to the pitch (intensity of noise). Sometimes, the intensity of the noise is so undesirable that is not liked by any one.

Hazards of Noise Pollution

Investigations in USA, Britain and Russia have revealed that noise pollution not only affects directly the hearing mechanism of animals including human beings but also tells upon their overall health. It hinders bodily efficiency by causing mental upset, insomnia and irritability leading to palpitation of the heart, rise in blood pressure and general debility. The hazardous consequences of air pollution and water pollution are easy to see but noise pollution does not leave much evidence that can be seen and identified immediately. It, however, leaves lasting effects, some short and some long on our body mechanism and metabolism (Gupta, 1999).

It has been found recently that 90 to 110 million people are bothered by environmental noise pollution everyday and approximately 50 million are adversely affected in terms of health in India. Big cities are much more afflicted with noise pollution than the smaller ones. Contrary to this, the remote places of Himalayas are the ideal centres of quietness and peace. In bigger cities more than 20 per cent people feel difficulty in having sound sleep whereas the rural folk in the remote corners of the country enjoy a sound sleep.

According to recent report published in *New York Times*, there are some strange and noxious effects of noise pollution, both to human beings and animals. Some of them are as follows:

Effects on Human Beings

Noise has been found to interfere with our activities at five levels:

(i) At the audiological level by interfering with the satisfactory functioning of the hearing mechanism.
(ii) At the biological level by interfering with the biological functioning of the body.
(iii) At the psychological level by interfering with the sleeping mechanism and functioning of muscles.
(iv) At the behavioural level by affecting the social behaviour of the subjects.
(v) At the personological level by influencing the personality.

Thus, noise pollution influences the performance of audiology, biology, psychology, behavioural and personology of the human beings.

Audiological Effects

Studies indicate that short exposure to high levels of noise may result in temporary deafness–lasting from a few seconds to few hours. A noise of 90 dB for more than 10 seconds contracts the tympanic membrane through aural reflex. Prolonged exposure during working hours can produce a permanent incurable deafness. The permanent deafness is caused when the noise exceeds 120 dB and less damage is caused if the noise is between 90 to 120 dB. The intensity of damage to hearing power, however, depends upon:

(a) The nature of noise,
(b) The intensity of noise,
(c) The place of noise,
(d) The nature of the noise producing instrument, and
(e) Duration of exposure.

Noise capable of causing permanent damage to hearing mechanism are commonly those originating from explosion, jet aircrafts and certain industrial processes in which the noise is amplified by diaphragm action. Explosive sound of about 150 dB can rupture the ear drum causing permanent damage to the nerve endings and associated parts of cochlea. Noise of 130 dB can cause tickling sensation or pain in the ear (Table 4.2).

Noise intensity of 79-90 dB spread over 8 hours exposure daily shatters the auditory system. This also leads to severe disturbances even affecting memory power and powers of retention and expression. It has also been observed that when a person is exposed for a long time to noise levels above 50 dB his fine hearing abilities not only get impaired for good but even person's work efficiency goes down.

Auditory fatigue is another diseases of its kind due to high intensity of noise. It may produce whistling and buzzling sounds in one's ear which implies that one may be unable to distinguish certain sounds when other sounds are present.

Biological Effects

The adverse effect of noise pollution on human beings are also manifested through various biological indications. It disturbs the nervous system, prevents concentration, brings tension and fatigues, interferes with metabolism and causes a number of other diseases.

Unchecked exposer to noise can cause afflictions ranging from headaches, instability in mental and physical equilibrium to insomnia, mental tension and makes a man short tempered. Noise between 60 and 120 dB causes mental stress as well as loss of work efficiency.

Medical evidence suggests that noise can cause heart attacks in individuals with cardiac injury and continued exposure to loud noise can cause chronic effects as hypertension or ulcers. In fact, quicker heart beats, shrinking of blood carrying and blood taking vessels, raising of blood pressure are the results of continuous exposure to high intensity sounds ultimately producing heart afflictions. On the effect of noise on the cardiovascular system, Dr. D.P. Agarwal (All India Institute of Medical Sciences 2000), cited a Japanese experiment in which 300 workers were exposed at work to sound level of 85 to 120 dB.

A slowing of heart rate was observed among them. Besides, at these noise levels, blood vessels constrict, pupil dilate, voluntary involuntary muscles become tense.

Table 4.2: Effect of different levels of noise on human ear

Level of Noise	Effects
Zero dB	Threshold of hearing
25 dB	General Silence
30-55 dB	Ordinary Conversation
60 dB	Normal conversation
65 dB	Does not pollute the environment
75 dB	Moderately loud and not so uncomfortable.
85 dB	Uncomfortable
85 dB (8 hours exposure)	Impairs hearing ability of the worker.
95 dB	Very loud and more uncomfortable.
110 dB (constant exposure everyday)	Develops hearing problem
120 dB	Severe pain and temporary loss of hearing.
130 dB	Tickling sensation or pain in ear.
150 dB or more (Few seconds)	Severe pain and temporary loss of hearing.
150 dB (constant exposure)	Deafness
150 dB (Explosive sound)	Rupture the ear drum

Noise pollution can change man's physiological state by speeding up the pulse and respiratory rates. Sound noises interfere with the supply of oxygen and removing wastes from the body.

The W.H.O. report has highlighted the impact of excessive sound on women particularly those in the state of pregnancy. Menstrual bleeding due to noise pollution is generally prolonged and lasts longer than usual. The worst effect of noise pollution has been found on the growing foetus and it can cause premature child birth. It is also believed that noise pollution retards sexual feelings. In places where high intensity noise keeps prevailing it can lead to psychiatric indisposition and can be a great hazard to infants. Highly painful noises seriously interfere with reproductive system causing vomiting , pain and restlessness in women apart from much bleeding during menstrual cycle. Loud noise hastens the appearance of the first menses in urban girls earlier than in their rural sisters.

In Japan it has been found that at high level of sounds (85 to 120 dB) the gastric juice in stomach and salivary secretions are decreased resulting in upsetting of digestive system. Adrenalin is suddenly injected into the blood stream, increasing neuro muscular tension, nervousness, irritability and anxiety. In epileptic noise brings on cunvulsions.

Psychological Effects

Psycho-physiologically unwanted noise creates communication disruption, frustration, sleeplessness, besides lack of co-operation and social conflicts.

Negative mental consequences include paranoia and suicidal and homicidal tendencies. A man's visual reaction drops by 25 per cent, when noise levels reach 90 dB. Noise is also said to cause a narrowing of pupils, after colour perception and reduce night vision.

Symptoms of psychosis, fearful and indignant dispositions inner and outer are some of the symptoms of sound intensities which are disturbing.

Intolerable agony may result when the source of the noise is not known. Interruptions in speech communications may impair performance, lead to errors and lower output and efficiency. Noise can cause tension in muscles, nervous irritability and strain.

Behavioural Effects

Many behavioural charts are recorded as a result of exposure to high level noise in human beings. Certain symptoms can be noticed outrightly. The undesired sound may cause annoyance. By lowering down the auditory sensitivity of person, noise results in poor attention and concentration. It has been observed that performance of school children is poor in comprehension whose schools are situated in busy areas of cities, suffering from noise pollution. Noise causes irritation which results in learning disabilities. Sudden noise distracts a person and can create nervousness in him. Housewives working in kitchen with all kinds of gadgets feel headache and giddiness due to noise and vibrations of these gadgets.

Inability to think, analyse and solve problems, accumulate tension, uneasiness to settle down and indifference to environs, are some of the symptoms caused due to certain abnormalities in behaviour, which get added up slowly but certainly during noise pollution.

Long and continuous exposures to sounds of very high intensities tell upon functioning of systems and this may affect growth, result into dwindling of thinking faculties, and failure to respond such things in the system, so essential for integrating various functional processes, also interfering with co-ordination between different system in the body.

Personological Effects

When the injurious effects of noise tend to persist for long, they cause stable maladaptive reactions in an individual, disturbing his total personality make up. In children, it may develop a feeling of inadequacy, lack of confidence, and poor perception of one's own self which may jeopardise their optimal personological development as a growing child. Once these feelings are developed by a child in the growing age, its disastrous effects are not going to be removed easily without leaving their marks behind.

Other Effects

Other effects of noise pollution are summarised as under:

(i) Noise has other physical effects also. According to medical researchers that loud noise sometimes causes sudden blindness, stammering and even fits of epilepsy – particularly in children.

(ii) A British scientific team has some evidence to show that noise of jet engines contributes to the breakdown in mental health of people living near jet airports. The team surveyed two groups of people, one living within the maximum noise area surrounding the London airport, and the other in an adjacent quieter zone. The team further found that admission to the mental hospital was 20 per cent higher from the noiser area. The affected people were living near the airport were mostly women. They suffered with neurotic illness.

(iii) High intensity of noise causes abortion in women and hormonal blood levels are seriously disturbed. Buzzing sensations and cyanosis are also the consequences of noise.

(iv) The older people are easily disturbed by the highly unpleasant and loudness. Their sleep is interfered which older people want more. High levels of environmental noise can disrupt the sleep of the babies also.

(v) The unwanted sound can influence unborn babies producing malformation of the foetus nervous system that may affect behaviour later in life.

(vi) Noise can contribute to a cause of gastrointestinal problem such as peptic ulcer.

(vii) The noise interferes with the peace of mind and creates discomfort.

(viii) Loud noise causes a number of physical disorders, emotional tantrums and other chronic effects.

(ix) Noise produced by vehicular horns near the hospitals or schools is a great nuisance. The patients get disturbed and feel uncomfortable, while in the institutions the blowing of horns disturb the students in attaining their academic pursuit.

(x) A sudden loud noise can set off a chain reaction. The muscles grow tense. There is imbalanced functioning of various system, the veins shrink, the heart beats faster and the blood pressure shoots up.

(xi) Industrial noise harms the workers much more than ordinary city noise. Since, it is constant, the mill workers are known to feel numbness around their ears. Child labourers suffer from permanent hearing losses due to their exposure of debilitating noise level at tender age.

(xii) Noise increases the possibilities of errors in both skilled and manual labour as well as mental work, the chances of accidents occurring is considerable and safety of the operators and others is at stake.

Effects on Animals/Birds

Not only human beings, but several animals and birds are affected by noise pollution. The animals like guinea pigs, rats, dogs and cats are subjected to intense noise levels over relatively short periods of time in place of low levels over long time periods for studying the effects of noise. The researchers have been able to conclude from these findings regarding damage caused to these animals. These damage include impairment of hearing mechanism, blood chemistry and damage of brain, heart and liver. Most of the farm animals show good adjustment to noise particularly horse, swine and cattle.

Noise pollution also leads to mental tiredness in animals. Damage to bone, joints, swellings and stiffness are also the result of noise pollution. Animals are driven into convulsion by sudden loud sounds, although there is no evidence that human beings are as susceptible to noise as animals are. In experiments conducted on rats exposed to noise levels above 500 dB, were found to have lost their responses and behaved erratically.

A number of birds are found to stop laying eggs under high intensity sounds. Moreover, a number of birds change their places from noise polluted areas. There is decline of migratory birds to a place when it is subjected to noise pollution.

Most of the poultry birds cannot adapt to loud sounds, particularly to unexpected loud noise. In fact, one of the earliest recorded official complaints related to aircraft noise occurred in 1923 when a farmer wrote to the Postmaster General complaining that flying aircraft were disrupting egg production.

Effects on Buildings

It has been reported that high intensity sounds even affect the building severely. Under the stress of unpleasant and loud sounds there is development of cracks and fissures in the buildings and other structures. There is also break down of window panes with high intensity sounds.

Managing of Noise Pollution

Based on three basic principles of noise control viz, reduction of vibrating sources, enclosure of the source and attention to the absorption of the sound, the following measures can bring down the volume of noise within the tolerance limits or even cut it out altogether (Gupta, 1999).

Creation of Vegetation Buffer Zones

Vegetation buffer zones of various evergreen trees and shrubs act as excellent sources of sound absorption. So these must be created within the densely populated areas of cities and big towns, around the schools, hospitals and offices. It has been observed that a dense evergreen hedges can reduce the noise by 10 dB during garbage collection in the towns.

Strip Plantation

A strip of 6 m wide plantation of trees inside the compound wall, effectively protects the house from the noise pollution created by the vehicular traffic.

Planting of Rows of Tall Trees

Planting of tall trees in rows, bushes and shrubs with ample foliage on both sides of busy roads serve as noise absorbers because they function as noise filter. Planting of trees particularly *Polyalthia longifolia* and *Azadirachta indica* whose leaves act as noise absorbers, is useful to control noise.

Recently a number of plant species were identified by Srivastava (2001), which are found suitable for absorbing noise and keep the atmosphere tolerably free from noise. They are shown in Table 4.3.

Use of Carpets/Drapery

Inside the buildings spreading of carpets/draperies and acoustical files provide reduction of noise levels, although such internal treatments do not affect the noise intrusion from outside.

Provision of Gaskets

Gaskets around doors of the house can also reduce noise transmission from one place to another.

Sealing of Outlets

Noise intrusion into the building can best be checked by sealing all the outlets to the outside. In extreme noise pollution cases, double structures can be employed. They can reduce the noise level by about 5 to 10 dB for high frequency sounds.

Checking Noise Source

Excessive noise can be prevented by checking various noise sources *viz,* radios, televisions, tape-recorders, stereos, while turning up these audio visuals, one must think over others.

Use of Spongy Material

If a soft spongy material is placed on the walls, floors and ceilings, the reflected sound is diffused and absorbed.

Table 4.3 : Noise Absorbent Plant Species

S.No.	Local Name	Botanical Name
1.	Neem	*Azadirachta indica*
2.	Jangal Jalebi	*Inga duleis*
3.	Semal	*Bombax ceiba*
4.	Bargad/Borh	*Ficus bengalensis*
5.	Peepal	*Ficus religiosa*
6.	Katrer or Kachnar	*Bauhinia variegata*
7.	Pink mohar	*Cassia nodosa*
8.	Not known	*Cassia racemosa*
9.	Amaltas or Karangal	*Cassia fistula*
10.	Pangara tree	*Erythrina variegata Var Orientalis*
11.	Rumbal	*Ficus racemosa or Ficus glomerata*
12.	Mahua	*Madhuca latifolia*
13.	Ashoka	*Saraca indica*
14.	Nili Gulmohar	*Jacaranda mimosaefolia*
15.	Panjtara or Jerul	*Lagrestromia flosregia*
16.	Gulmohar	*Delonix regia*
17.	Palah or Dhak	*Butea monosperma*
18.	Sal	*Shorea robusta*

Source: Srivastava, Ramji (2001). Personal Communication. Forest Research Institute, Dehra Dun.
Sharma, P.D. (1998). Ecology & Environment.
Rastogi Publishers, Shivaji Road, Meerut-20002, India

Better Designs

In case of machinery, automobiles, motor cycles and aircrafts, noise control can be achieved by better design and effective silencing and sound proofing devices and measures at the manufacturing stage.

Exacting Standards

Vehicle horns must be subjected to the most exacting standards of noise control. Their indiscriminate use should be avoided and silence zones declared in cities around educational institutions, hospitals and offices.

Setting up Noise Limits

Noise pollution can be overcome by strictly enforcing the Air Pollution Act and setting up of noise limits in the urban areas. Penal action against the violators of the law should be taken.

Restrictions to Use Old Vehicle and Ban on Use of Horns

The Pollution Control Board must identify the sources of noise pollution and action should be taken to reduce them. Old vehicles should not be allowed to ply as they cause maximum noise. Similarly traffic police should take stringent action to force motor vehicle users to cut down, noise emitting from their vehicles to minimum. For this silencer should be in good condition. Ear piercing horns should be banned as is the case in western countries.

Noise pollution in fact can be controlled through legal restrictions.

Engineering Techniques

Noise exposure reduction may be achieved by the application of engineering control techniques or by the regulation of exposed personnel.

Engineering control techniques refer to the alteration of design, changes in the operation of noise source or construction of sound barriers and sound absorbers. Regulation of exposed personnel means supply of air protective devices or lowering of their exposure time.

Substitution of Quieter Equipment

Noise from equipment may be reduced by substitution of quieter equipment or quieter process activities. Examples are the use of centrifugal fan in place of axial flow fan, the use of welding to replace a noiser rivetting process.

Public Awareness

Public awareness about noise pollution and its hazardous impacts through electronic media can also go a long way to reduce noise pollution and make the environment healthy.

It has also been reported by World Health Organisation that all environmental problems like air pollution and water pollution, noise pollution is the easiest to overcome. What is needed, is to create awareness and mass awakening to the hazards of the noise pollution.

Noise Reducing Technology

Some of the recent aircrafts such as the 747 have incorporated noise reducing technology developed in last few years. Sound absorbing things of intake and exchange can reduce noise, but of course with some penalties of added weight and expanse.

Use of Mufflers

A stream of air and some moving fluid can produce excessive sound levels due to turbulence established in the ambient atmosphere. In many situations the use of a muffler is the most convenient method of noise control particularly the trucks and motor cycles. A dissipative muffler installed in a gas flow pipe makes use of absorbing materials to reduce sound for horn or fibres may be used as sound absorbing material.

Use of Ear-muffs/Ear Plugs

In noise generating factories, the workers should be provided with ear-muffs or ear plugs and wool to cut out piercing harmful noise for protecting their ears. In fact, these should be provided along with tickets to travelling public also (on demand) in railway and buses.

Cotton wool with a dab of vaseline is a cheap protection and most effective method to control noise pollution.

Rules and Restrictions

Control must be exercised on crackers during festivals, other ceremonies, celebrations and rallies. The enforcement of Anti-pollution laws should be strictly enforced by the law enforcement agencies. Furthermore, use of loud speakers should not be allowed except for certain important occasions. Indeed the use of permanent outdoor loud speakers in any religious or non-religious place should be totally banned. They should only be permitted to fix them inside the halls or interiors, for the benefit of worshippers or devotees.

Checking Canine Population

The canine population both domestic and wild also plays a significant role in causing noise pollution particularly during nights. Domestic population of dogs and cats must be kept under control. Acceptable noise levels are given is Table 4.4.

Table 4.4 : Acceptable Noise Levels (dB)

Residential	Bed room	25
	Living room	40
Commercial	Office	35-45
	Conference	40-45
	Restaurants	40-60
Industrial	Workshop	40-60
	Laboratory	40-50
Educational	Classroom	30-40
	Library	35-40
Hospitals	Wards	20-25

Source: Koenigsberger, O.H. (1973). *Manual of Tropical Housing and Building*. Part-I. Climatic Design, Orient Longman, Bombay.

Suggestions

- Organise environmental weeks to make the people familiar with adverse effects of noise population with NGOs, religious leaders and public.
- Involve the college and school students as well as local people to create awareness regarding noise pollution.
- For efficient enforcement of noise pollution laws, it is best to empower the Directorate of Environment.
- Planting of tree/bushes be encouraged in the noise pollution prone areas.
- Since noise is directly proportional to congestion of buildings and

pollution, Urban planting should be stressed on dispersal of settlements.

- Pollution Control Board must identify the sources of noise pollution in greater detail and action to reduce them, should be taken on priority.
- The traffic police should take stringent action to force motor vehicle users to cut down noise emitting from their vehicles to the minimum. Ear piercing horns should be banned.
- There should be regular driving course for the drivers in which, among other things blowing of horns unnecessarily should receive proper attention. Similarly, the district authorities should first educate the people on laws related to noise pollution and then try to enforce these laws.
- As prolonged exposure during working period produces permanent deafness which is incurable. Therefore, it is suggested that in highly noisy work places, workers should be taken off to an area of quiet for short intervals by building some noise free rooms.
- Noisy aircrafts like Boeings should be grounded until noise suppressors are fitted. Additionally, all air lines operating through Indian airports should be asked to use the low drag, low power landing approach successfully employed by the German's Lufthansa Airline, to cut airport noise by half.
- No noise abatement programmes can succeed without people's participation. It is, therefore, suggested that their education through all available media should be undertaken on priority basis.
- Although it is difficult to reduce harmful noise significantly in the environment of cities, it is easier to do so in our homes. Though we may not be able to create a perfectly quiet and peaceful atmosphere around us. Yet by consistent efforts at home we make condition more tolerable by way of reducing the sounds of T.V., radios and transistors.
- Industrial areas should be built well outside the city and workers given medical check-up periodically.
- A campaign should be launched throughout the country to educate people to the effects of noise pollution and dangers arising from it.
- Noise abatement programmes should be prepared and launched on a national scale.

- Unless the government and every individual does something to cut down noise, the 90 years old prediction of Robert Koch, a Nobel Laureate and Bacteriologist, may soon come true, "A day will come when man will have to fight merciless noise as the worst enemy of health".

Difference between Air/Water and Noise Pollution

Noise pollution is different from other types of pollutions. It kills us mentally, inflicting us with all sorts of mental ailments, deaf and dumbness, restive and disturbed condition.

As with inhaling polluted air and drinking polluted water the ill effects begin appearing after some time, so is the case with noise pollution. If this menace is not checked now, most Indian will end up having severely impaired hearing faculties by the turn of the century. Presently, it has been investigated that three out of five Indian (especially in big cities and towns) have already lost some power of hearing which they will never retrieve.

Noise Pollution – Legal Prospectives

The Government of India adopted very deep and sincere policies and planning during 1970 to provide clean environment through environmental management. This is very clear from the National Fourth Five Year Plan (1969-74) which highlighted the following environmental issues.

(i) It is an obligation of each generation to maintain the productive capacity of land, air water and wildlife in a way which leaves its successors some choice in creating healthy environment.

(ii) The physical environment *viz.* land, air and water is a dynamic, complex and interconnected system. Action in one of these components affects the other. There is also independence of living things (Animals and plants) and their relationships with land, air and water.

(iii) Planting for harmonious development recognises its unity of nature and man. Such planning is possible only on the basis of comprehensive appraisal environmental issues particularly economic and ecological.

(iv) There are instances in which timely, specialised advice on environment aspects could have helped in project design and in averting subsequent adverse affects on the environment, leading to loss of invested resources. It is, therefore, essential to introduce the environment aspect into our planning and development alongwith effective conservation and rational use of natural resources, protection and improvement is vital for national well-being.

The environmental matters got a special significance in the Sixth Five Year (1980-1985) Plan with the objective to conserve environment and its control. Perceiving the problems of environment and its impact on national development, environmental management has come to occupy a place of priority in the hands of Government. Now a separate Department of Environment (DoE) has not only been created at the centre, but also in each and every State of India to tackle the ecological crisis and environmental problems.

Legal Control of Noise

Various kinds of pollution (air, water, noise and soil pollution) are assuming alarming proportions because people are still ignorant about their hazardous consequences. These are destroying the eco-system and consuming the qualities of the environment. Hence it becomes imperative to regulate the various types of pollutions through certain laws, so that the environment is saved from being further polluted in the interest of mankind.

A number of countries have enacted specific legislations to control noise pollution, including England and America. In England there is Noise Abatement Act of 1960 in which Section 2 provides that (a) loud speakers shall not be operated between the hours of nine in the evening and eight in the morning for any purpose, (b) at any other time for purpose of advertising and entertainment, trade or business. Of course some exceptions are prescribed in the Act. But in India, there is no such Act, which exclusively deals with problems of noise and its control. As a result the drivers go on blowing horns unnecessarily whether they need it or not. If their vehicles are stranded in a big traffic jam then what is the purpose of blowing horns loudly ? It is just a habit with them.

The US, Noise Pollution and Abatement Act, 1970 is an important legislation for controlling noise. Under this Act the Environmental Protection Agency acting through the office of Noise Abatement and Control, holds public meeting in selected cities to compile information on noise pollution. Although in India, there is no separate law for regulation noise pollution, yet the Indian Constitution embodies provision in Article 39 (e), 47, 48-A and 51-A (6). Article (e) states that the health and strength of workers, men and women, and the tender aged children are not abused. Article 47 enjoys upon the State to raise the level of nutrition and standard of living and to improve public health. Since 1976, Article 48-A provides:

> "The State will endeavour to protect and improve the environment and to safeguard the forest and wild life of the country. In certain States viz. M.P., Rajasthan and Bihar, there are Acts which provide restrictions against the use and play of loud speakers".

The Environmental Protection Act, 1986 which came into force from 19th November, 1986 to commemorate the memory of Mrs. Indira Gandhi who was globally recognised as a dedicated environmentalist. This act purports to inculcate environmental ethics in every citizen and to take appropriate steps for the protection and improvement of the environment. Section 2 of the Act defines "environment to include air, water and land the interrelationship with living organisms (Animals and Plants)".

Impact Points to Remember for Noise Pollution

- Noise is unwanted sound. It is becoming more and more prominent as a pollutant.
- With the increasing industrialisation of our society, noise is becoming slow poison.
- Thunder, perhaps the loudest natural sound we hear, it sometimes reaches the threshold of discomfort.
- People experience excessive noise from television, desert cooler,

generator, exhaust fan, electric grinder, washing machine, hand pump installed with motor.

- People feel disturbed from neighbouring environment like aircraft take off, traffic noise, loud speakers, horns, construction work, railway lines and various kinds of industries.
- Heavy industries like iron and steel production fabricating and mining, construction of industries often expose their employees to hazardous noise levels.
- Variation in noise levels is wide. Noise is measured in dB.
- Slowly but surely, noise is becoming an unjustifiable interference and imposition upon human comfort, health and quality of modern life.
- The drivers of passenger cars often are less bothered by their own noise than their fellow drivers.
- In rural areas, ambient noise can be as low as 30 dB even in residential areas in or around cities, this level is seldom achieved.
- In urban areas, the noise level can be 70 dB or higher for 18 hours each day.
- There is sufficient evidence that noise beyond a certain level causes hearing damage, raises blood pressure and cardiovascular diseases like tachycardia (high pulse rate), errors in work and understanding speech.
- Noise also causes low birth rate, affects immune system and increase in psychiatric admission to hospitals.
- Noise leads to stress and digestive disorders.
- Workers exposed to high noise levels had higher incidence of ear-nose-throat (ENT) problems and cardiovascular disorders.
- To solve noise pollution, one must find out something about what the noise is doing, where it comes from, how it travels, and what can be done about it.
- The best way to control noise is to develop a green belt vegetation at regular intervals within densely populated areas of cities and towns.
- A strip of 6 metres plantation inside the compound wall of the houses and planting of rows of tall trees/bushes/shrubs with ample foliage on both sides of the highways should be developed.

- Trees, shrubs and bushes act as noise absorbers and noise filters.
- Inside buildings sound absorbing materials such as carpets, drapery and acoustical files should be used to reduce noise pollution/intrusion.
- Gaskets around door of the houses can also reduce noise transmission.
- Excessive noise can be prevented by checking noise source such as radio, television, tape recorder.
- The persons employed at machines must be provided with ear muffs, and cotton/wool plugs to protect their ears.

REFERENCES

Agarwal, D.P. (2000). *Effect of Noise on the Cardiovascular System on Industrial Workers,* All India Institute of Medical Sciences, New Delhi (Personal Communication).

Anonymous (1996). From our Jammu Diary. *United India Pride.* Monday, November 18,1996.

Gupta, R.D. (1999). Noise Pollution and its management. *The Kashmir Times. Sunday Magazine* October 24th, 1999 issue.

Koenigsberger, O.H. (1973). Manual of Tropical housing and Building. Part-I, *Climatic Design,* Orient Longman Bombay.

Sharma, P.D. (1998). *Ecology and Environment.* Rastogi Publishers, Shivaji Road, Meerut, U.P. India.

Srivastava, Ramji (2001). Noise Absorbant Plant Species. Forest Research Institute, Dehradun, U.P. India (Personal Communication).

5
VEHICULAR POLLUTION

How to Define Vehicular Pollution

Vehicular pollution may be defined as the product of incomplete combustion of fuel in the form of exhaust emission. According to chemical reaction the product of complete combustion of air-fuel mixture should be carbon dioxide, water vapour and nitrogen. But due to complex nature of reaction of the combustion process and other factors such as type and nature of engine, fuel quality, other systems and lubricants incomplete combustion takes place. This results into the liberation and emission of lethal compounds like carbon monoxide, unburnt hydrocarbons and nitrogen oxides in the atmosphere from petrol fueled vehicles, and nitrogen oxides, sulphur dioxide and suspended particulate matter from diesel fueled vehicles.

Status of Vehicular Pollution in India

In our country, vehicular pollution has increased significantly during the past few years. This problem is, however, more serious in metropolitan cities. Highly alarming issue pertaining to this is the sudden spurt in the number of vehicles which shot up to 9.17 millions in 1986 to 25.30 millions in 1993, recording a rise of about 16 millions in seven years (Suri, 1999). Further, there is an additional number of 2.3 million vehicles every year. So, what is the total number of vehicles plying on the roads now would be anybody's guess. Most of these vehicles are concentrated significantly in urban areas particularly in metros and State capitals. In Delhi city alone, 17 lakhs vehicles were registered in 1986 which shot up to 22 lakhs in 1994 (Suri, 1999) and 28 lakhs in 2000 (Sunder, 2000). The number of vehicles plying daily in Delhi is around 30 lakhs. Again, in Delhi alone, the number of vehicles is more than the combined total in Mumbai, Kolkata and Chennai.

As per the published records (1993-1994) vehicles emitted about 1300 tonnes of pollutants in the air of Delhi everyday as compared to 870 tonnes in 1987 (Suri, 1999). Thus, in a span of 7 years the vehicular pollution load increased by nearly 50 per cent in Delhi. Significantly air pollution load in Delhi is more than sum of vehicular pollutants in Mumbai (640 tonnes day^{-1}), Kolkata (311 tonnes day^{-1}) and Bangalore (254 tonnes day^{-1}). It is noteworthy that out of the total air pollution load from various sources "Vehicular Emission" comprises a major contributor. Shockingly, 64 per cent of air pollution in Delhi and 52 per cent in Mumbai owe its origin to vehicular emission. As the air grows perceptibly darker and smoggy, it affects eyes, ears and respiratory system. Although industries, thermal power plants and domestic activities also contribute sizeably to air pollution in urban areas, yet it is the vehicles or transport sector which is the largest emitter of carbon monoxides, nitrogen oxides and hydrocarbons in the air. Gasoline run vehicles contribute most of carbon monoxide, hydrocarbon and lead, while diesel vehicles are the chief source of particulate matter and sulphur dioxide. About two-third of air pollution in Jammu city is due to vehicular emission (J&K State Pollution Control Board, 2000).

Factors Responsible for Vehicular Pollution

In accordance with the study conducted by Centre for Science and Environment in various metros, and other capital cities in India in the era of vehicular pollution indicated significant findings. It was noted that Vehicular Pollution is attributed broadly to four main factors:

(i) *Bad vehicular technology* – An outdated technology.

(ii) *Poor fuel quality* – From the source and adulteration at distribution outlets. In India, this is a real problem.

(iii) *Poor vehicular maintenance* – We always try to economise on maintenance.

(iv) *Non-existent traffic planning* – There are frequent stoppages on account of many intersection and traffic congestion which lead to excessive emission. Moreover, it is not enough to fine tune individual vehicles, it is ever more important to keep a strict control on the total population of vehicles in a city.

(v) *In addition to these factors*, adulteration of fuel and sale of spurious auto components are cited as two major hurdles in ensuring optimum operative performance of vehicles. These are the deep rooted problems which most of the private and public vehicle owners do not have control.

Legal Implications for Vehicular Pollution

In the interest of every human being at large, and to keep the environment neat and clean, and to make air and water free pollution, various Health Organisations in the world have become alert. As a result certain restrictions and regulations have been framed for strict implementation. Some important provisions in regulations to control vehicular pollution caused by Transport Sector, are briefly mentioned as under:

- To keep vehicular pollution under control, the Government of India has framed certain rules relating to emission of smoke from every motor vehicle and have set up some norms which all vehicles have to fulfil. They are incorporated in the Central Motor Vehicle Rules (CMVR) of 1989 and are amended from time to time. There are separate rules and norms for Petrol and Diesel driven vehicles which have to be strictly enforced and complied with by all as per amended CMVR No. 115 Sub Rule (2). Every Motor Vehicle shall comply with the prescribed standards such as at idling speed of the vehicle the carbon monoxide emission limit for all 4 wheel petrol driven vehicles shall not exceed 3.0 per cent by volume. Such limit for 2 and 3 wheeled petrol driven vehicles shall not exceed 4.5 per cent by volume. The maximum smoke density for all diesel driven vehicles shall be upto 5.2 Bosch Units (65 to 75 Hartidge units).
- As per Sub Rule No. 7 of CMV Rule No. 115 as well as Rule 116 every motor vehicle shall carry with them all time a valid

"Pollution Under Control" (PUC) Certificate issued by the agency authorised for this purpose by the State Government and shall have to be produced on demand by any checking officer. The said certificate can be issued only after conducting the test in prescribed manner to measure the standards of emission of smoke from the vehicles. The certificate shall have to be renewed from time to time.

As per Sub Rule of CMV Rule No. 116, the measurement of the smoke emission for compliance of various provisions shall be done with a metre of the type approved by various Central Government agencies like Vehicle Research and Development Establishment (VRDE) of Ministry of Defence, Government of India, *i.e.* VRDE, Ahmednagar.

Thus, standards for controlling emission level of new vehicles were incorporated into the Motor Vehicles Act (1989) only in 1991 enforcing some sort of quality control on Automobile industry. The Bureau of Indian Standard (BIS) for the fuel started incorporating emission parameters almost during the same period. However, in both these cases, the standards were set according to terms and conditions dictated by Auto Industry. As a sequel, standards followed today are too lax as compared to followed elsewhere in the world.

Hazards of Vehicular Pollution

Public Health studies have shown that with higher levels of Vehicular pollution there is heavy rates of bronchitis, asthma and acute respiratory infections particularly in children. Besides, cardio-respiratory diseases, kidney disorders, hypertension, diabetes and many other diseases, including skin ailments have been found relating to higher vehicular pollution in the air. The vehicular pollution level in the country's capital is to such an extent that the incidence of respiratory diseases is about 12 times the national average. Above all, about 30 per cent of Delhi's population suffers from respiratory diseases due to vehicular pollution.

Briefly, the effects of vehicular pollutants on the health of human beings are shown in Table 5.1.

Table 5.1: Effects of Vehicular Pollutants

Pollutant	Effects
CO (From gasoline cars, 2 and 3 wheelers)	Fatal in large doses, aggravates heart disorders, affects central nervous system, impairs oxygen carrying capacity of blood.
NO (from diesel vehicles)	Causes irritation of respiratory tract.
Ozone	Produces eye, nose and throat irritation, risk asthmatics especially in children and those involved in heavy exercise.
HC (mainly from 2 and 3 wheelers)	Causes drowsiness, eye irritation and coughing.
Pb (from petrol vehicles)	Extremely toxic effects, nervous system and blood, can impair mental development of children, causes hypertension.
Benzene	Carcinogenic.
Aldehydes	Irritation of eye, nose and throat, sneezing, coughing difficulty in breathing, carcinogenic in animals.
Polycyclic aromatic hydrocarbons	These are mainly responsible for producing cancers.

Source: Tumbde and Malewar (2000). *Environment and People* 7(4) : 14-15.

Managing of Vehicular Pollution

Following steps should be taken immediately to improve the governance of Vehicular Pollution.

To Promote Non-motorised Transport

It is the time when we should seriously promote non-motorised transport like the bicycles and battery or solar energy driven eco-friendly vehicles. The cyclist must be encouraged.

To Improve Fuel Quality

As fuel and engine go hand in hand, so we must improve fuel's quality first. Without better fuel we cannot introduce better technology. The Government must take steps to improve fuel quality because this is the only short term measure which can ameliorate air quality considerably.

To Promote Non-vehicular Mass

Promotion of non-vehicular mass transit systems is very essential, so that large proportion of passenger trips can be undertaken on the electric trains or trams.

To set on Vehicular Pollution Control Standards

Vehicular pollution control standards must be set on health related criteria and not on the basis of what is technically feasible for the automotive or petroleum industry, as it is at present.

To set up National Vehicular Pollution Prevention Commission

A national Vehicular Pollution Prevention Commission should be set up by 'Government of India'. Similarly vehicular pollution control commission should be set up by the State Government at the level of each city facing vehicular pollution problems.

To make Available Decisions to Public

All Government documents related to decisions making on pollution control must be made available to the public on demand.

To Restrict Vehicular Population

Government should frame certain rules to restrict vehicular number. Everybody should not be allowed to have the vehicles but certain criteria must be fixed prior to purchasing the vehicles.

To Promote Vehicles (Cars, Jeeps) Pooling

Promoting pooling of cars, jeeps and other vehicles by employees of Corporations, Factories, Institutions and Universities is another way to decrease the number of vehicular pollution.

Use of Catalytic Converters should be encouraged which does not have much cost (Kaul, 1998)

Suggestions

An awareness campaign must be launched in big cities with the objective to reduce vehicular pollution by educating people about the health hazards, caused due to vehicular pollution.

Free facilities should be provided to vehicle owners at petrol pumps in the city to enable them to get their vehicles tested/checked.

The vehicle owners and drivers of polluting vehicles should be given diagnosis reports and advised to get these defects rectified to reduce pollution levels. "Paryavaran Vahini Scheme" should be commenced as has been started in a number of cities of U.P. with the objectives to monitor and check pollution caused by vehicular emissions. For this purpose, pollution control testing kits must be provided to each of the Paryavaran Vahini. The members of Vahini should also disseminate relevant literature and information to generate enthusiasm among citizens, and involve them in environmental issues.

The drivers of the vehicles whosoever violates the traffic law should be awarded deterent punishment.

Impact Points to Remember for Vehicular Pollution

Vehicular pollution in India is reaching at an alarming levels especially in mega cities like Kolkata, Chennai, Mumbai, Delhi, Ahmedabad, Bangalore and many others.

The increasing number of vehicular traffic coupled with setting up of industrial units the Jammu city and its outskirts have become agog with environmental pollution, which if not checked well in time may lead to disastrous consequences (Sharma, 2000).

In Jammu region, about 1500 to 2000 vehicles are added every month which works out to be about 20,000 (Sharma, 2000) per annum. As our roads infrastructures are not adequate to contain this recurring complexities resulting in occasional traffic jams.

Pollution caused by the public and private vehicles is a big hazard to living beings. All human beings, animals, trees and monuments and other buildings get affected by the smoke emitted by petrol and diesel vehicles.

Lead emitted in the atmosphere from auto exhaust in micron sizes remains suspended in air and enters respiratory system with inhalation process, causing lead poisoning of the human body leading to stomach ailments, indigestion, constipation, lack of appetite and sleeplessness.

The vehicular emission is the cause of air pollution, global warming, erosion of ozone layer, fog and various health related problems.

The smoke and other gases emitted from the engines pollute the atmospheric airgases like hydrocarbon and oxides of nitrogen emitted from automobiles are the major sources of photo chemical smoke which affects the human body badly.

The pollution from the vehicles is caused by nature and quality of fuel, engine, design, efficiency and maintenance, traffic conditions and weather conditions.

Use of better refined petroleum products is called for. Appropriate legal action against oil companies or petrol pump dealers supplying adulterated quality of fuel requires to be taken seriously and promptly.

Every vehicle should be checked for pollution parameters and it should be confirmed with prescribed standards before being registered for plying on roads. Periodical checks of vehicles for ensuring safety standards is very essential. Vehicles emitting visible smoke should be impounded on spot and made to rectify the defect within a day or so.

Poisonous emissions from vehicles increase with erratic movements, it is high at low and very high speeds and minimum at traffic at idle running. It is necessary to regulate the traffic.

Roads require to be well maintained, made wide and free from pot holes and surface distortions.

Speed breaker, narrow roads, dumping of sand, metal chips along roads, traffic jams, dusty and garbage filled roads increase vehicular pollution.

It needs to be recalled that engine using excessive fuel not only wastes precious national resource but also causes greater pollution.

Trees and green patches have a good effect in absorbing

vehicular pollution. Hence, planting of trees must be encouraged along the road sides.

Stop import of two wheelers and three wheelers till their engines are made four stroke. Introduce only fuel efficient vehicles.

In the long run we shall have to bring in battery operated or CNG driven vehicles which are almost pollution free.

REFERENCES

Koul, V.K. (1998). Cleaning the air. *Popular Sciences* 6 (1) : 37-38.

Sharma, G. (2000). Growing Vehicular Pollution dragging Jammuites to death trap. *Daily Excelsior*. Sunday Magazine, October 22, 2000.

Sunder, I. (2000). Problems and Challenges of Sustainable Development in India. *Environment and People*. 7(6) : 27.

Suri, B.K. (1999). Increasing Vehicular Pollution. *Daily Excelsior*. Thursday, August 5, 1999 issue.

Tumbde, M.M. and Malewar, M.M. (2000). Pollution under control check alone may not put brakes on vehicular air pollution. *Environment and People* 7(4): 14-17.

6
METAL POLLUTION

The incidence of air and water pollution from metals has reached such an alarming levels that environmentalists are finding it difficult to enforce control measures. The salts of heavy metals such as Hg, Pb, Cd and many others constitute poisons. These are in fact known as third kind of pollutants, the others two are non-degradable and biodegradable pollutants. All these pollutants seriously interfere with bio-environmental processes, and thereby, posing a menace to the life on this planet (Gupta, 2000).

Metals have been classified into essential and non-essential groups. The former consists of Ca, Mg, K, Fe, Mn, Zn, Cu and the latter of Hg, Pb, Cd, Cr, As and Se and plus some more. The metals of the second group have directly or indirectly adverse consequences on biological activities. The presence of these metals in water and air degrade their qualities which eventually damage the life of human population in one or the other way. Even the essential metals at higher concentration than the permissible limits, prove harmful. The present chapter deals with sources, the hazards of metals *vis-à-vis* their management.

Sources of Metal Pollution

The main sources of metal pollution are as follows:

Various Industrial Processes

Electro plating/plating wastes, discharges from smelter operations and batteries, as well as manufacturing of pigments and refining and welding processes aggravate the pollution potential of cadmium in the air. The estuary of the Hudson river in Canada had been so affected, causing a very real danger to inhabitants nearby. Consumption of plant produce grown at such sites is also considered unsafe.

Air borne zinc poses a threat to life which mainly occurs in the form of zinc oxide (ZnO) with coloured vapours. From zinc refineries, some zinc enters the air besides from other sources such as brass and bronze refineries. Zinc as byproduct from copper refining units also enters the air.

Automobile Industries

The second important polluting agent is lead, particularly from automobile industries. The use of lead tetraethyl as an anti-knock compound (in place of certain blends of diesel and petrol which can be effective) has increased the problem in recent years. Unfortunately, the Indian automobile industry does not follow the Western technological model which is less hazardous and more environment friendly.

Industrial Water

Waste water from industries as well as incinerations are other causes of metal pollution in many countries. Switzerland and Japan are worth mentioning examples in this respect. When water is used in hydroelectric projects, its only energy is utilized and when it is used in industries as coolant, its only temperature is altered. However, when water is utilized in chemical industries it becomes contaminated with acids, phenols, polyvinyl chloride and metal like mercury and chromium. There are a number of well documented cases regarding mercury discharges into different waterways. Hitherto it was not thought that mercury would undergo alterations when introduced into water to form methyl mercury compound. But now, it is believed that this compound enters the food chain in lower orders of the plants and then advances into the marine food chain and ultimately to our tables. The Minimata incident in Japan is probably the best known example which affected the fish population in the area and later on the human population. Mercury is not the only element but there are others too *viz.* copper, cadmium, lead, selenium and arsenic, which adversely affect the human bodies.

Wastes from Pulp and Paper Industries

According to experts, mercury pollution also poses the threat to the environment largely through the pulp and paper industries which discharge the metal into the stream and rivers. In absence of suitable wastage disposal system, mercury poisoning is a possibility and Indian can no longer wish away. The pollution potential of mercury is more than three times than that of lead worldwide.

Exhausts, Automobiles, Gasoline

Lead enters into the air from exhausts automobile, vehicle fumes and particulate matter. Gasoline possessed tetraethyle lead. Lead affects longevity and its poisoning can do a havoc. The lead poisoning was perhaps noticed in New York in children during 1954-1967. In accordance with, Dr. Veena Kalra, Paediatrics Professor (All India Institute of Medical Sciences, New Delhi, 1998), the presence of lead in the blood of children from two schools each at Daryaganj and Asiad Village (Delhi) was attributed to the vehicle fumes and dust inhaled by the children.

Coal Mine/Dust/Fly Ash/Cement Plants

Toxic trace elements present in coal mine dusts is another serious source of metal pollution. Fly ash generated from thermal plant and dust of cement plants are the other sources which impart metal pollution at an alarming rate and posing health hazards.

Rocks and Minerals

Rocks and minerals contain various elements which on weathering are carried away by the running water into streams and rivers. Dr. Nitish Priyadarshi, fellow member Geological Society of India and life member of Indian Association of Sedimentology, said that the sediments carried away by a rivers like Damodar and the Ganga were being deposited in parts of West Bengal including Burdwan where various disorders and diseases caused by the deadly arsenic have been reported. Presence of considerable amount of arsenic in coal mine areas (Bachara Piprawar) of South Bihar has considerably affected the water of the Damodar and its tributaries.

Contamination of Ground Water

Other factors responsible for the arsenic contamination of ground water in parts of West Bengal are geological and not anthropogenic. Study carried put by Central Ground Water Board (1999) revealed that more than 5 million people nearly 30 per cent of them children in eight districts of West Bengal, are severely exposed to arsenic menace. It was due to high concentration of the metal in ground water. Actually the arsenic problem first surfaced in West Bengal in the late eighties. Subsequently, this problem was also noticed in Bangladesh and is now being reported from Orissa. In a cluster of villages in Khurda tehsil of Orissa where people are being crippled due to high levels of fluoride in drinking water besides arsenic. According to one study, about 44 million people in India are affected by arsenic as well as excessive fluoride and nitrate.

Soil Acidity

Ulrich Forestner of the Technische University Hemburg in West Germany found that acidity of soil increases the mobility of metals. It also influences the chemical equilibria and unless suitable measures are taken, to control these factors, metal pollution levels will reach a critical point within two years.

Organic Wastes

Direct contact of garbages with soils as well as unscientific practices such as composting and fertilizing and indiscriminate dumping of wastes are other sources of metallic pollution. These are also the source of soil pathogen, disease bearing flies, attack of ants and rats, viable worm eggs and viruses. Heavy metals like Cd, Cr, Ni etc. accumulate in soil over a period of time to the toxic level. These elements finally get accumulated in human beings and animals through agricultural/horticultural/olericultural crops which are grown on such soils, and thus create a number of complicacies.

Hazards of Metal Pollution

The pollution of environment including metal pollution is the gift of industrial revolution, which has adversely affected the ecological balance and caused serious health hazards. Problems of pollution by metals have aggravated for a variety of reasons, the most common being the unpredictability of mobility. For instance, chromium perturbations are mostly localized but hexavalent and trivalent chromium behave differently. It has been observed that aluminium is less problematic in comparison to lead, chromium, mercury and cadmium. Indeed, cadmium is highly toxic and inhibits the functioning of certain enzymes. The gastro intestinal tract and lungs are most susceptible to cadmium poisoning. It has been estimated that the average biological half life of the metal is 17.6 years and a daily intake of 200 μg can damage the kidneys of human beings having age greater than 50 years. In US and Canada, the daily intake of the metal has been found to range from 50 to 80 microgram through food chain. Cadmium accumulation leads to suppression of growth, bone deformations, renal problems and bloodlessness. It also causes anaemia and atrophy in some parts of reproductive systems.

Mercury is another contaminant released by using mercury containing fungicides, paints and cosmetics. Lethal levels of mercury are not exactly known so far. However, less than 0.1 mg/m^3 mercury is permissible. Mercury deposits lead to headache, pain in some parts of body, intestinal problems as diarrhoea, blood malfunctioning.

Chromium is said to be carcinogenic agent. It also leads to some kidney disorders, ulcers in stomach and in intestines and some disorders in nervous system.

Lead affects the longevity and cause poisoning in biological systems. According to Basel based centre for disease control, 10 microgram/decilitre of lead in blood is considered normal. Higher levels (> 40 microgram/decilitre) can lead to anaemia, abdominal pain, damage to nerves, convulsion and hypertension and even prove lethal, says Dr. Veena Kalra who has found blood lead levels in 200 school children in the national capital. The natural growth of child could be hampered if he has high dose of lead in his system. The red blood cells lose their haemoglobin binding capacity.

Arsenic leads to liver damage, ulcers and some kidney problems. The symptoms of chronic arsenic poisoning include various types of dermatological disorders, muscular weakness, liver complications and paralysis of lower limbs.

Copper has its own effects. It leads to mental stress and disturbance, coma and tension. It also results in uremia, a disturbance when blood cannot dispose of some poisonous matter deposited in it. Selenium accumulates in plants grown on selenium rich soils.

Accumulation of zinc produces kidney problems, pain in legs and even vomiting. A brief account of different diseases caused due to intake of various metal pollutants are shown in Table 6.1.

Though traces of fluoride in the food are essential especially for reducing the incidence of dental decay, Recently, a number of fluoride pockets have been located by the geologists which impart high levels of fluoride to the drinking water. This causes fluorosis a bone disease prevalent mainly in A.P., Rajasthan, Punjab, Haryana, U.P. and Jammu and Kashmir (Sharma *et al.,* 2000).

The cumulative effects of aluminium are now receiving serious attention because of its wide use in cooking utensils. High levels of aluminium in human body weakens our nervous system and as such is very harmful. All metals beyond certain limits are toxic (Table 6.2).

Managing of Metallic Pollution

In light of the aforesaid factors amenable to disturb the ecological balance and posing threats to mankind and other organisms, the following methods are proposed for managing metallic pollution.

Plantation of Trees

Maximum thrust should be applied to plantation of trees for creating ecological balance especially in the mining/quarrying degraded areas. This will not only prevent the pollution but also adds to the scenic beauty and aesthetic value of such areas. While planting various tree species emphasis should be given to the local species, and exotics must be discouraged.

Table 6.1 : Diseases caused by metal pollution

Metal Pollutant	Effects and diseases
Cadmium	Suppression of growth, bone deformation, renal problems, anaemia, liver injuries, gastric and intestinal disorders, problems in function of central nervous system.
Mercury	Headaches, pain in some parts of body, intestinal problems as diarrhoea, blood malfunctioning, minimata disease characterised by numbness and vision disorders.
Lead	Mild anaemia, brain damage, vomiting, loss of appetite, convulsions, uncoordinated body movements eventually producing coma and death.
Chromium	Carcinogenic agent, kidney disorders, ulcer in stomach and intestines, some disorders in nervous system.
Arsenic	Vomitting, diarrhoea, nausea, severe irritation of nose and throat, abdominal pain, skin eruptions, inflammations and even death, liver damage, ulcers, kidney problems.
Zinc	Kidney problems, pain in legs, cramps, fever, vomiting, renal damage and pneumonitis.
Copper	Mental stress and disturbed, tension and coma, uremia and even death, sporadic fever, pathological changes in brain tissue.
Selenium	Agent for fever, problems in teeth, blood pressure fall, problems in spleen and liver, blindness.
Iron	Rapid increase in respiration, pulse rates, congestion of blood vessels, drowsiness paralysis, diarrhoea, low blood pressure, long irritation, bone defects.
Manganese	Growth retardation, fever, sexual impotence, muscular fatigue and eye blindness.
Aluminium	Gastro-intestinal, irritation and shaver's disease in lungs.
Barium	Intestinal malfunctioning, salivation, gastric problems, vomiting.
Calcium	Depresses the functioning of muscles and nervous tissues, hyper calciemia causes coma and death.
Cobalt	Some problems of bones, blood pressure and fluctuations paralysis.

Source: Gupta, R.D. (2000). *Environment and People*. 7(4) : 10-12.

Monitoring the Levels of Toxic Elements

Methods evolved for monitoring levels of toxic elements in the air, should be strictly followed. These method consists of using ultraviolet radiations in exchange resins and ultrafiltration.

Table 6.2 : Permissible limits of certain elements

Name of the element	Permissible limit (ppm)
Nickel	0.02
Manganese	0.5
Lead	0.01
Arsenic	0.01
Selenium	0.01
Chromium	0.05
Cadmium	0.003
Barium	0.7
Iron	0.05
Fluoride	1.5
Copper	2.0
Zinc	< 5.00
Mercury	0.001
Calcium	< 200
Magnesium	< 150
Molybdenum	0.07

Source: W.H.O. (1993) As per report of Park's Text Book of Preventive and Social Medicine, 1997, Jabalpur, India.

The methods devised by the scientists of the National Environmental Engineering Research Institute (NEERI) at Nagpur, should also be used to destroy hexavalent chromium.

Treating the Organic Wastes

Leading steel companies have outlined measures for treating wastes containing hexavalent chromium with ferrous sulphate. These measures provide for facilitating discharges of treated water in permissible concentrations into the surface water. Sewage and sludge, constituting a potent source of metal (Tables 6.3 & 6.4) must also be treated.

Table 6.3 : Heavy Metals in Sewage Sludge

Metal	No. of sites	Amount of heavy metal (microgram^{-L})	
		Range	Mean
Arsenic	10	6 – 230	43
Barium	60	21 – 4980	576
Cadmium	115	8 – 846	101
Chromium	119	17 – 99000	3280
Mercury	53	1 – 10600	1077
Nickel	109	10 – 3515	440
Lead	116	13 – 19730	1656

Source: Cited from Panwar *et al*., (1988). Intensive use of Agrochemicals and Heavy Metals – A Threat to Environment Pollution. *Farmers and Parliament*. August issue.

Table 6.4: Heavy metal concentration in sewage sludges of different basins of Dal Lake complex in ppm

Dal Lake basin	No. of samples tested	Cd	Zn	Cu	Pb	Ni	Cr
Boddal	16	32	1240	685	384	86	85
Gagribal	16	43	1185	714	412	80	104
Hazratbal	16	48	1094	854	425	108	112
Nagin	16	21	1248	654	210	68	96

Source: Goverdhan, V. (1989). Heavy Metal Pollution of different inland lakes of Northern Himalaya, *Indian J. Environmental Protection* 9(7): 531-536.

Use of Special Culture

A special culture called rhizobium has proved useful to deal with manganese ore for restoring soil productivity and fertility in the region of waste dumps.

Organizing of Seminars/Symposia

Local administration should conduct seminars, symposia, debates to create metallic pollution awareness amongst the urban and rural masses. There must be close liaison among people and government officers/officials working in the line.

Use of Metal Substitutes

Use of metal substitutes like wood and glass must be encouraged. Wood and glass can also be used for many articles now made of metals. Use of plastics can also conserve metals.

Government Laws

The Government laws should be strictly implemented to protect the environment and pollution control.

Suggestions

- Extensive programme of research by the department of Science and Technology should be undertaken to check the metallic pollution being caused by mining and quarrying.
- Some suitable local agencies must be established for conservation and monitoring of metallic pollution.
- Reviewing of laws and ordinances pertaining to mineral resources responsible for creating metallic pollution and updating them in accordance with consistent regulations and insights that will allow multiple land-use of areas necessary for mineral extraction.
- The encouragement of additional research aimed at providing new technologies for ore recovery and improvement to lessen the metallic pollution.
- Setting up of tubewells tapping arsenic free (aquifer) ground water instead of upper arsenic rich aquifer. It also stressed the need of development of community based water quality surveillance and monitoring system.
- Creating of awareness in the community by alerting them about the menace of arsenic poisoning through drinking water as well as other metals. They should be enlightened that arsenic dermatosis is not a contagious disease and can be avoided when arsenic free water is taken.

Impact Points to Remember for Metal Pollution

- A number of metals enter the human systems through air and water pollutions.
- The metallic pollutants are present in pesticides, organic wastes and inorganic constituents.
- The soil is the primary recipient and from the soil, the metallic pollutants make their entry into the living beings.
- Of the various metals, Cd and As are extremely toxic while Hg, Pb, Ni and F are moderately poisonous. On the other hand, B, Cu, Mn and Zn are relatively lower in toxicity.
- In fact all metals are toxic when present beyond permissible limits.
- Lead pollution in India and other South Asian countries comes from battery recycling units, lead mining, soldering industries and old water supply pipe lines.
- Lead levels in the air over Dhaka and Mexico, is about 8 times higher than the safe level as declared by WHO.
- In seven cities of India *viz*. Mumbai, Delhi, Chennai, Hyderabad, Bangalore, Kolkata and Vellore, more than 50 per cent of the children below 12 years age had Pb content beyond permissible limit in blood.
- Fluoride related health hazard is a major environmental problem in many regions of the world.
- Recently large areas of Assam, have been found fluorosis endemic.
- In many parts of the country especially eastern region, the underground water is polluted with Fe and Mn.
- Cadmium being more mobile in soil and more by plants than other heavy metals, it has more potential to move from plant to man and other animals.
- High levels of Cd have been found in vegetable crops grown in road side soils.
- Heavy metal pollution causes a number of diseases to human beings. For example Cd accumulation leads to bone deformation,

Cromium is a carcinogenic agent and Cu leads to mental stress.

- Maximum thrust should be given to plantation of trees for preventing the pollution.
- Methods evolved for monitoring levels of toxic elements in the air should be strictly followed.

REFERENCES

Goverdhan, V. (1989). Heavy Metal Pollution of different inland lakes of Northern Himalayas. *Indian J. Environmental Protection* 9(7) : 531-536.

Gupta, R.D. (2000). Metal Pollution – hazards, sources and management. *Environment and People*. 7(4): 10-12.

Kalra, Veena, (1998). Presence of lead in the blood of children from two village schools of Delhi. (Personal communication) AIIMS, New Delhi.

Keeney, D.R. (1982). "Nitrogen management for maximum efficiency and mini pollution". In F. J. Stevenson (Ed.), *Nitrogen in Agri. Soils,* Am. Soc. of Agronomy, Madison, Wisconsin, pp. 605-649.

Panwar, B.S.; Chaudhary, M.L. and Karwasta, S.P.S. (1988). Intensive use of Agrochemicals and Heavy Metals – A threat to environment pollution. *Farmer's and Parliament*. August issue pp. 27-28, 35.

Sharma, S.; Vora, J.J.; Marfani, M.F. and Joshi, J.D. (2000). Fluoride Problem in North - West. *India, Environment and People* 7(7):7-14.

7
DUST POLLUTION

Introduction

Pollutants are of diverse types which may be present in the air, water, soil and food. They may be present in very small quantity or in greater amount depending upon the source of contaminants. Pollutants may be in the solid, liquid or gaseous form. Air pollutants not only occur from smoke, noxious gases, fumes, garbages, nightsoil but also from the dust.

Dust pollution in our country has increased significantly during the past few years. The problem has been more serious in towns or areas where stone crusher units are present largely. The places or sites having cement factories and mines and quarries are also not free from dust pollution. Deforested and denuded areas are more liable to have dust pollution. The presence of dust in the atmosphere not only degrades air but also the water and soil. A thick layer of dust on green trees along the roads is an index of dusty environment. Not only this, lawns, parks, gardens, markets, public places, offices and libraries also appear to be dusty. A visible layer of dust is seen every morning on all household articles, furnishing and even on edible stuffs left uncovered. Dust cleaning is a tedious and tiring job for the house wife. Valuable textile articles sometimes get spoiled permanently due to dust. Books are worst affected.

Causes of Dust Pollution

The causes of dust pollution are well known and, therefore, solution of this problem should be easy.

Disappearance of Vegetation

Disappearing vegetation and denuding lands in our environment are the main causes responsible for creating dusty atmosphere or dust pollution. Jammu has become a most dusty city due to denudation of the Siwalik hills. Similar is the condition of the Siwalik hills of Punjab and Himachal Pradesh, Morni hills of Haryana, responsible for creating dusty atmosphere of a number of towns especially during summer months (May and June) and autumn months (October and November). Denuded hills of Giri and Bata rivers catchments in Himachal Pradesh, have shown sheet, gully and landslide/slip erosion besides dust pollution (Sharma *et al.*, 1998) like Siwaliks of Jammu, Punjab and Himachal Pradesh.

Thus degradation of forest and vegetation from the Siwaliks and other hills to distant places around Jammu and other towns and ever increasing area of bare soils is the cause of spreading dust all over.

Maintenance of Roads

Dust clouds following vehicles on the dusty roads which have not been repaired for years, can be cited as one of the spectacular example for dust pollution. These clouds spread dust all over the cities and suburbs, and make over life miserable. Nightsoil from open defecation, as it dries, is blown away to contaminate air, water, food and other edibles. Dust is thus source of infection.

City dust, thus, comes from the roads neglected for long time. On papers, these roads must be shown repaired but on ground there is not a piece of metal or *bajri* seen on these roads. Dust clouds following vehicles is common phenomenon while en routing from Jammu to Surian Sar and Mansar roads, Patnitop to Sanasar and Chenani to Sudh Mahadev and Mantalli. Similarly while moving from Andretta to Chadiar Road and from Paunta Sahib to Shalli, Shimla to Kotkhai in Himachal Pradesh, dense dust clouds can be seen.

Winds

Dust pollution is responsible for causing marked rise in total suspended particulate (TSP) which is a measure of the concentration of all particles in the air tiny or big. In Delhi alone TSP levels show a marked rise in summer months of May and June because of dust from the winds, with a sharp decline in rainy months of July, August and September because the rains wash off the dust particles in the air, and then a sharp rise again in October. The winter TSP levels from October to February are generally high but not as high as in the summer months of May and June. But even in the monsoon months, the TSP levels stay well above the permissible limits. Although annual average levels of TSP have fluctuated over the years yet they have consistently remained above the national annual average standard of 140 $\mu g\ m^2$ from 3.85 times the limits in 1987 and 2.15 times in 1998. The Central Pollution Control Board has described air quality as critical when TSP is 1.5 times the standard. The cities and towns adjoining to Thar desert are too afflicted with TSP particularly during summer.

Dust pollution is also very common in cold arid region of Ladakh in Jammu and Kashmir, and cold arid zone of Lahaul, Spiti and Kinnaur in Himachal Pradesh, due to blowing of heavy winds and extreme aridity. The area of Marwah block of Kishtwar tehsil, Doda district adjoining to Ladakh region also suffers badly with dust pollution because of blowing of winds with high speed every day during summer (Rana and Bindra, 1980).

Mining and Quarrying

Although minerals both non-metallic (Limestone, dolomitic limestone, gypsum) and metallic ones (Bauxite, iron, copper and zinc ores, gold and other kinds of ores) are important natural resources of any country from economy point of view. However, unscrupulous unscientific mining, often using excessive explosives, on steep slopes in Mussoorie hills and Sirmour and Kashmir Himalayas have done immense damage to fragile ecosystems of these areas besides producing dust pollution. The mass erosion from the mined areas of Mussoorie was to the tune of 550 tonnes ha^{-1} (Dadhwal *et al.*, 1998). Degradation of land, drying of water resources, disruption of communications, floods and consequent decline in food and milk production are the other ill effects, which have been manifested due to unscientific mining (Juyal *et al.*, 1995).

Owing to accumulation of the mine spoil/debris, the characteristics of the soils have totally changed (Dadhwal *et al.*, 1998). Due to presence of high gravel content and less fine material the infiltration rate of the spoil becomes high and water holding capacity low. The debris is alkaline and calcareous in nature.

While extracting the metals and minerals, a number of mining operating *viz.* crushing, powdering, concentration of the ore; and calcinations are done in addition to blasting of rocks. All these mining operations emit a large quantity of dust, sand, silica and fluorine in the atmosphere.

People working in the coal mines, sulphur mines, lead mines and stone mines inhale a variety of gases which cause a number of respiratory diseases. The labourers working in the said mines/rocks are often subjected to silicosis disease and those working in coal mines die of carbon compounds, labourers working in the salt mines often become bald and those working in fluorine mines die due to fluorosis disease.

The gypsum mining in Kishtwar, the bauxite mining in Reasi and quarrying in Ramban of Jammu region and lignite mining in Kashmir, have all changed the environmental scenario of the Jammu and Kashmir State.

Cement Factories

The growth of cement factories is polluting the atmosphere unabated. The effect of cement factory dust has been seen in Kashmir in the diseased crop of saffron, fruit trees and other tree species.

The effect of cement kiln dust on the growth and development of *Shorea robusta* forests around Rajban area (near Paunta Sahib indicated (Gupta *et al.*, 1992) that :

(i) There was change in soil reaction from acidic (pH 5.7) to (basic pH 8.5).
(ii) Sulphur content of the soil increased.
(iii) A significant decrease in chlorophyll and starch content of the leaves was found to reduce the photosynthetic efficiency of the plants.

Stone Crushers

The crusher units produce disastrous effect on environment in the form of dust and noise pollution. The recent study conducted by Semban and Chandrasekhar (2000) with regard to the impact of dust pollution on the health of the crusher workers and found that:

(i) The workers are severely suffering from a number of skin and respiratory diseases (Table 7.1).
(ii) The workers are taking the treatment from Government and private hospitals and hence their expenses on diseases have increased.

Health Hazards of Dust Pollution

Exposure to dust can produce four distinct types of disabilities (Ahmad, 1993). These are pneumoconioses, silicosis, asbestosis and mental fume fever, which are caused only by dust inhalation. Metal fume fever is a temporary malady resulting from inhaling certain mettallic oxide fumes. Dust carries spores of fungi, eggs of insects and also many resistant disease causing germs which sometimes becomes causes of some epidemics.

Allergy is a common problem of mankind which is caused by dust and dust blown pollens and other infectious elements in the air. The allergic reaction or protein poisoning, as typified by hay fever, is the direct result of breathing pollen and other organic substances. Headache, chest pain, fever, eye and skin diseases, lung infection and diarrhoea are the other ailments which are producted by the dust pollution.

Management of Dust Pollution

Vegetative Measures

Since disappearing vegetation and denuding lands in our environment are responsible for dust pollution, so covering of every inch of the bare land with vegetation-trees, shrubs, herbs, grasses and turfs-is the solution. Every owner of land, government and private, must take responsibility of keeping his land under vegetation cover.

Reforestation and control of wind erosion is the only remedy. While reestablishing a vegetation cover over bare hills a long term process, we can help the situation to a great extent if our roads, parks, gardens, lawns, play grounds and public places are maintained in good condition.

Table 7.1: Classification of Age, Education and Diseases caused by Dust Pollution for Stone Crushers

Education			Age-wise (years)			Disease	
Different Level	No. of workers	%	Different level	No. of workers	%	Various diseases	No. of workers
Illiterate	44	41.5	upto 18	15	14.1	TB	12
Primary	37	34.9	18-36	56	52.8	Eye & Skin	44
Secondary	24	22.6	36-54	30	28.3	Asthma	14
Technical	1	0.9	54-72	5	4.7	Diarrhoea	35
Total	106	100	-	106	100	Others Nill	106

Source : Semban T., and Chandrasekhar, S. (2000). *Environment & People*. 7(7): 29.

As the climate of cold arid regions is very harsh, marked by sharp temperature fluctuations, extreme aridity, dusty atmosphere, low oxygen tension, wind and snow storms, so if forest cover is increased particularly in desertic wastelands it would reduce temperature fluctuations, dust and effects of wind and snow storms. On the whole, the climate will become healthier and salubrious (Sharma and Mir, 2000). Experiments have shown that close forest of one hectare can filter nearly 50 tonnes of dust particles, thereby, protecting human beings from the adverse effects of dust pollution.

In areas subject to sand movement due to wind erosion, should be combated through establishment of shelter belts. *Salix* and *Populus* species which give a bushy growth under Leh conditions have been thriving well near water courses and should be planted vigorously. They will not only ameliorate the environment but also protect farm land and provide herbage, fuel wood and poles for buildings.

Mechanical Measures

In highly degraded mined areas, establishment of vegetation is difficult due to high run off/debris movement, lack of moisture and absence of fertile soil. Mechanical or engineering measures aptly called as first line of defence become imperative at such areas to protect the planted vegetation by arresting debris and create conditions conducive to plant growth by accumulating fine soil and improve moisture status.

Mechanical measures adopted in the mined watersheds mainly consisted of:

- **(i)** ***Grade stabilization structures*:** In the main drainage channels a series of gabion (wire crates filled with stones) cross barriers, silt detention basins and checkdams are constructed to arrest the debris and reduce the bed gradient for safe runoff disposal.
- **(ii)** ***Slope stabilization structures*:** The mine spoils/dumps consisting of mineral rejects and overburden in the mined areas are highly unstable and easily erodable by runoff. Mild slopes are stabilized by digging contour trenches and planted with suitable species like *Vitex negundo, Saccharum spontaneum, Acacia catechu.*

Vegetative measures are also taken alongwith engineering structures. The plant species with high root binding characteristics which could survive in the poor sites, are preferred. Grasses such as *Crysopogo, fulvus, Pennisetum spp. Saccharum* spp. *Eulaliopsis binata,* are the best proposition in this regard. Among the various trees *Leucaena leucocephala, Acacia catechu, Cedrela toona, Bauhinia* spp. have been found very suitable for the purpose.

Impact Points to Remember for Dust Pollution

- Dust pollution is a source of infection. Dust carries spores of fungi and bacteria and many disease causing germs.
- A thick layer of dust on green trees along the roads is an index of dusty environment.
- Photosynthetic activity of some of the plants having a thick layer of dust is reduced and thereby growth is impaired.

- Protection against dust is an uphill task and dust proofing is an impossibility.
- Our most sensitive electronic devices like computers, television sets, tape recorders are badly damaged by dust.
- City dust comes from the roads neglected for long time. These must be repaired frequently.
- The labourers working in the stone crushers/mines and quarries, are afflicted with a number of diseases like tuberculosis, asthma, diarrhoea, eye and skin ailments, chest pain, fever, headache as well as silicosis, asbestosis and pneumoconioses.
- Mining's other impacts are drought and soil salinity, are making life impossible for farm settlements and pastoral families.
- Toxic dust metals like lead, manganese, and cadmium can provide their effect as a result of either breathing or swallowing dust.
- Total suspended particulate levels themselves are frightening, air pollution related public health experts are now arguing that not all particles are equally deadly. They become more and more deadly as they become smaller and smaller.
- Dust particles less than one micron are more harmful as they go deep into the lungs and stay there longer and are thereby able to cause the maximum damage to the lung. For this reason these particles are also called respirable particles.
- Disappearing vegetation and denuding lands in our environment are amenable for causing dust pollution.
- Covering of bare land with various species of trees, shrubs, herbs, grasses and turfs is the suitable method to control dust pollution.
- Reforestation is also essential for controlling wind erosion.
- Although afforestation/reforestation can control dust pollution and wind erosion to some extent yet nothing escapes dust since it is wind blown and air borne, and there can be no place without air and dust on the Earth.

REFERENCES

Ahmad, M.S. (1993). Kinds of Pollution. *Indian Council for Enviro-Legal Action.* 2 (3): 11-12.

Dadhwal, K.S.; Juyal, G.P.; Katiyar, V.S. and Joshie, P. (1998). *Rehabilitation of Degraded Mined Lands of North-West Himalayas.* Success stories on TOT, SAIC, BARC, Campus, Farmgate Dhaka 1215, Bangladesh.

Gupta, R.D.; Kanwar, B.S.; and Rana, K.S. (1992). Ecological considerations of the Sirmour Himalaya. In *Himalayan Environment Man and Economic Activities* (Ed. Raina J.L.) Pointer Publishers Jaipur 302003 (India).

Juyal, G.P.; Katiyar, V.S.; Dhadwal, K.S.; Joshie, P. and Arya, R.K. (1995) Reclamation of mine spoils on steep Himalayan hill slopes. In sustainable reconstruction of high land head water regions (Ed. by Singh, R.B. and Martin, J.H). *Proc.* 3rd *International Symp.* On Headwater Control. Oxford and IBH Pub. Co. Ltd. New Delhi.

Rana, P.S. and Bindra, A.S. (1980). A trip to an exotic horticultural paradise, *Agalumnus.* 8 (2) : 4-6.

Semban, T. and Chandrasekhar, S. (2000). Impact of crusher pollution on worker in Trichy. *Environment and People.* 7(7)29.

Sharma, J.P. and Mir, A. (2000). *Dynamics of Cold arid Agriculture.* Kalyani Publishers, Ludhiana, New Delhi-Noida (UP) – Hyderabad – Chennai – Calcutta – Cuttack.

Sharma, P.D.; Sharma, S.P.; Sharma, K.D. and Minhas, R.S. (1998). Status Report on catchments of Giri and Bata rivers in HP, Department of Soil Science, HPKV, Palampur-176062.

8
SOIL POLLUTION

Introduction

Soil sustains innumerable plants, animals and microbes alongwith plenty of mineral reserves. Chemists consider the soil as store house of plant nutrients, while to a geologist the soil is the means of ascertaining the age of the earth from a large number of isotopes present in it. The soil to a physicists is composed of solid, liquid and gaseous material. The soil to a limnologist is a dependable source of nutrition for the biota, and to an agriculturist, it is the uppermost crust of the earth which is capable of growing plant life and other forms of life. Above all, the soil is a living entity wherefrom originated all major ecological events of the past. However, soil/land pollution is the direct effect of chemical, unscientific use of organic wastes and improper drainage. These being important natural resources, their loss through erosion and pollution has to be prevented.

Causes and Hazards of Soil Pollution

Soil pollution stems from a number of sources like man made chemicals, the waste materials from mining or quarrying and their processes, indiscriminate disposal of solid waste etc. The ill effects and health hazards from these sources are mentioned below:

Use of Chemical Fertilizers

There is no doubt that application of various fertilizers like nitrogenous fertilizers, phosphatic and potassic fertilizers has increased the yield of different crops especially of high yielding varieties of wheat, rice and maize. But their continuous use has contaminated the ground water rendering it unfit for human consumption and that for the consumption of all other organisms.

Almost all nitrogenous fertilizers are soluble in water and thus nitrates move readily into the water. During heavy rains or irrigation, the applied fertilizers are washed away carrying nitrates into the streams or rivers and also into drainage water. The danger of nitrate pollution is much greater in sandy soils owing to high percolation capacity (Gupta, 2002).

High concentration of nitrate in drinking water is the cause of methaemoglobinamia (blue blood baby disease) which may prove fatal for infants. The susceptibility of infants to nitrate has been attributed to their high intake. This problem, however, is not serious in adults. Water containing as low as 65 ppm of nitrate is unsafe. This disease has not been reported from areas where the drinking water contains less than 45 ppm which is the permissible limit of this pollutant. Fodders like oats, maize stalks when raised on soils with heavy application of chemical fertilizers may cause health problems like degeneration of vascular tissues of lungs, heart and liver in animals. The critical limit of nitrate in this case is reported to be 100-150 ppm.

Phosphorus being immobile nutrient, cannot be leached out much from the soil. However, inorganic soils having less amount of iron and aluminium, it becomes mobile and can get leached out in ground water as soluble organic phosphorus. Moreover, it is also carried away into the lakes and ponds through sediments. Addition of phosphorus alongwith nitrate applied in excessive quantity in soil will enter lakes and ponds, and will intensify eutrophication which is harmful to fish and other aquatic life.

Unbalanced application of acidifying nitrogenous fertilizer decreases soil pH and lowers the availability of phosphorus to plants. This is happening in some soils of *kandi* belt of Jammu, where the continuous use of nitrogenous fertilizer (urea) has produced soil acidity (pH 5.2 –6.5) (Gupta *et al.*, 1992).

Use of nitrogenous fertilizer alone has also started, producing negative effects in various crops and proved disastrous in low pH soils. Example of Palampur (H.P.) provides good evidence where the yield of maize with application of 100 per cent N alone was found lesser even than control (Kanwar, 1999).

Continuous use of N, P_2O_5 and K_2O fertilizers lead to depletion of micronutrients – Zn, Fe, Cu, Mn, MO, B–which may influence plant, animal and human health. Another hazard associated with excessive use of nitrogenous fertilizers is the gaseous loss of nitrogen into the atmosphere. Release of nitrous oxide in the stratosphere depletes the ozone layer, and by absorbing certain wavelengths of infra red-light, increases global temperature and destabilizes climates. This could lead to changes in patterns, levels and risks of agricultural production. A rise in sea level would have grave consequences for low lying delta and estuarine regions.

In view of these dangers, greater effort is required to promote more efficient and less polluting use of nitrogenous fertilizers and use of alternative source of nitrogen e.g., crop residues, animal wastes, green manures, legumes in rotations and use of biofertilizers.

Further excessive use of nitrogen in soils, especially nitrates will affect adversely the action of nitrogen fixing bacteria.

Use of Pesticides

Another important agriculture input added to soil is pesticides. Pesticides are a group of agricultural chemicals which include weedicides, insecticides, nematicides, fungicides, antibiotics and harmones.

Large number of pesticide compounds have a mobility class value of 1 indicating their high immobility. This groups includes compounds such as phorate, parathion, ethion, zineb, benomyl, paraquat, heptachlor, endrin, aldrin, chlordane, toxaphene and DDT. Some of these are already banned for use in agriculture.

The commonly used herbicides such as atrazine, alachlor, propachlor, simazine, propanil and diuron have moderate to high immobility indicating greater persistence in soil.

When the pesticides are used repeatedly in each crop season and those applied to soils which are poor in organic matter and microbial biodiversity, the chemicals are bound to accumulate for longer periods of time causing much soil and environmental pollution. Some of the herbicides like atrazine, benulide and chloroxuron have persistence in soil for periods ranging from 300 to 700 days under temperate conditions. The compounds with higher mobility in soil like 2, 4-D, picloram and dalapon may be leached out in water. This will result in much soil pollution. The relative mobility of pesticides in soils is as follows :

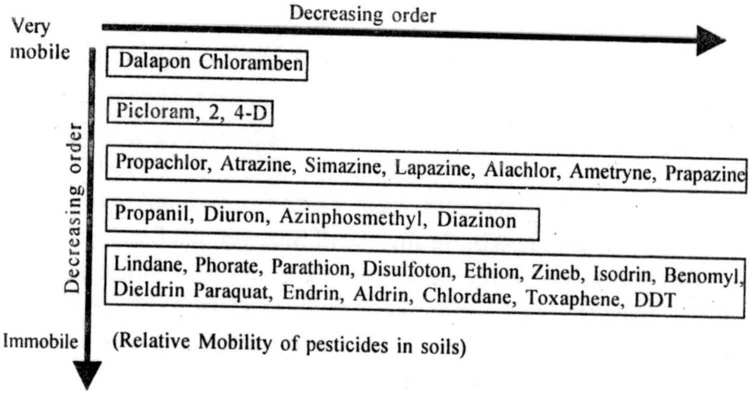

(Relative Mobility of pesticides in soils)

Pesticides reduce the population of soil organisms (micro and macro) which affect the soil structure and its fertility. Several pesticides and their degradation products are absorbed by the plants, which in turn may influence the entire food chains and food webs. Owing to slow absorption and accumulation of some of these chemicals (DDT, dieldrin, lindane) by lower organisms like water microbes and plants coupled with their regular and increased application, leads to "Biological Magnification". This then becomes carcinogenic or tetratogenic at or after achieving certain critical concentration. There are alarming reports about the increase in level of such toxic products in milk and other food products. Besides, they also prove toxic to many useful birds and insects like honey bees.

Pesticides not only kill macro and micro organisms of soils but also cause damage to crops *vis-à-vis* natural useful organism which act as enemies of pests. The incidence of pest attacks and secondary pest attacks may increase after pesticides have killed the natural enemies *i.e.* surgence.

Only a small portion of the pesticides applied in fields, reaches the organisms that are supposed to be controlled. The major part reaches the air, water or soil where it has a damaging effect on living organisms. Aquatic organisms are particularly sensitive to pesticides.

Pesticides that do not break down easily are absorbed in the food chain and cause considerable damage to insects, insect consuming animals, prey birds, and ultimately human beings.

The reckless use of pesticides in Punjab and Haryana has not only polluted the soil, but has also given rise to brown hopper (a very common pest of rice crop) which was hitherto not known. Development of alternative farming technology encompassing use of a agricultrutal wastes like cowdung and green manures is the need of the day.

Unscientific Disposal of Organic Wastes

In developing countries like India, the major sources of soil pollution is the unscientific disposal of organic wastes of various kinds. In rural areas, the organic wastes of animals are converted into manures and used in the fields to increase their fertility. However, these wastes are stocked or dumped in open fields and the process of their conversion to manures is left to be completed by the nature. This practice not only creates soil pollution but also becomes a major cause of air pollution. During bacterial decomposition of these wastes a number of gases (Ammonia, oxides of nitrogen and others) escape into the air. There is heavy loss of nutrients during rains. Hence this practice which is very common in Jammu and Kashmir rural areas also, should be totally banned. The farmers should be advised to prepare manures of their available organic wastes in pits, these should be sufficiently decomposed prior to using in the fields.

Direct contact of garbages with soils, besides producing ammonia, nitrate and methane, also become a source of soil pathogens, diseases bearing flies, attack of ants and rats. The dumping sites invite crows, vultures and dogs to scatter the rubbish here and there. Similarly unscientific use of sewage sludges are a source of soil pollution. These may contain enough living viruses and viable worm eggs as well as harmful bacteria, actinomycetes, fungi and other organisms. These when used in crop raising, which may be those eaten fresh both by humans and animals, are hazardous for health and environment. Similarly heavy metals like cadmium, chromium, nickel etc. accumulate in soils over a period of time to toxic level. These elements finally get accumulated in human beings and animals through agricultural products grown on such soils and thus create a number of complicacies.

Role of Salts

Similarly agricultural inorganic product become a source of soil pollution. To push up the animal weight and increase in milk production, has promoted the practice of increasing the soluble salt content in the animal fodders/fields. This has resulted in more salt content in the dungs of the animals when these dungs are used as manures they will contain higher salt concentration. The heavy and continuous application of such manures to soils may finally lead to increase of soluble salt content in the soils and in the ground water due to its leaching from the soils.

Increase in Salinity, Sodicity and Acidity

The land is being degraded by the increase in soluble salt (salinity) and sodium content (sodicity) due to water logging.

Increase in the concentration of soluble salts adversely affects the soils productivity and degrades the quality of land. Dissolved salts in irrigation water accumulate on the soil surface. This is aggravated by inadequate drainage particularly in flood prone areas and canal/well irrigated areas. In addition to this, the salts from the lower layers move up by capillary action during summer season and are deposited as white crusts on the surface.

Intensive farming with poor drainage is causing serious salinization in soils of India. Total area under saline and sodic soils is more than 7 million hectares. In Punjab alone an area about 6,000 to 8,000 hectares of land is becoming unfit for agriculture annually.

In Jammu district about 9,000 hectares of land has been found to be affected by salts due to uncontrolled use of irrigation water (Najjar, 1993). High salinity in soils of Kandi belt of Jammu is quite prevalent (Gupta *et al.*,1992) and also in soils of cold arid zone of Ladakh *vis-a-vis* valley of Kashmir (Gupta *et al.*, 1985).

Acid soils have an adverse effect on plant growth, which is

chiefly related to the presence of aluminium, manganese and iron in toxic amount. Low availability of calcium manganese and potassium, imbalance of micronutrients and low microbial activities also contribute to lower down the site quality. In Kandi belt of Jammu, the continuous use of urea alone has rendered soils of some localities acidic (Gupta *et al.*, 1992).

Land Degradation

Land and soil face several problems other than pollution. These are deforestation, overgrazing of the pastures, recurrence of floods, water logging and soil erosion. If soil depletion and land degradation continue at the current rates, it is estimated that about onethird of the total arable land will be destroyed by the end of this century.

Soil Erosion

Soil is eroded by water, wind, ocean, waves and glaciers. People's activities such as felling of trees, over-grazing of pastures, over-cropping and improper tilling, accelerate soil erosion. Erosion occurs in both wet and arid regions irrespective of modern or traditional agriculture.

Floods and soil erosion are caused by deforestation. Roots of grasses act as excellent binding material and keep the soil intact. Disruption of the grass cover by the plough, loosens the soil and makes it vulnerable for removal by the force of rains or strong winds. Tilling or grazing of slopes or semi-arid soils increase the rate of soil erosion. In India, deforestation and overgrazing have been the main causes of soil erosion. It is estimated that more than 6,000 million tonnes of top soil gets washed annually in India from 80 million hectares of arable land (Gupta, 2002). This figure almost becomes double when total culturable and unculturable land is taken into consideration. Most of this soil is washed into the sea, but some also collects behind dams, reducing their life span or gets deposited in river beds. The precious plant nutrient (N, P and K) that this soil (6,000 million tonnes) carries away are worth more than 700 crores of rupees. Every year the world's capacity to grow food, fuel and fibre is being reduced by the degradation of land and loss of soil.

Due to extensive cultivation of crops on hill slope and denuded forest area, soil itself can become a damaging agent through silting of fertile fields below. The suspended soils in runoff can reduce the sunlight penetration into water, thereby affecting organisms, which form the aquatic food chain.

The State of Jammu and Kashmir is most fragile as far as its geography and topography is concerned. The land resources of the State have been spoiled due to the deterioration of various ecosystems. With the present trend in population growth, the per capita availability of cultivable land has increased from 0.25 hectares to 0.16 hectares. The area under cultivation could be increased if we manage the available 1.64 lakh hectares of cultivable waste land, 8.92 lakh hectares of large chunks of eroded river beds degraded area and 11.78 lakh hectares of waste land, and miscellaneous 11.86 lakh hectares (Jamwal, 2002). These are the minimum figures, which the author has been able to obtain but may the more.

In Kathua and Jammu districts along the international Indo-Pak border there is some water logged area as well as in Kashmir particularly Sonawari area.

Acid Rains

As already mentioned in "Air Pollution" chapter that acid rains are the results of excessive quantities of sulphur dioxide and oxides of nitrogen in the atmosphere owing to industrialization.

The acid rain does a lot of harm not only to agricultural lands but is also harmful to forest resources, animal and wildlife, water resources and human health. Briefly its effects on various resources are recorded hereunder.

Agricultural Lands

The rise in the acidity in soils can bring a lot of chemical changes. Acidity in soils can release certain elements like iron, aluminium, manganese and some heavy metals like cadmium, mercury and other trace elements. In acidic soils, these elements are converted into more available forms to plants and as such cause toxic effect on their growth. Moreover, the availability of phosphorus, calcium, magnesium, boron and molybdenum to the plants is restricted due to which they suffer badly.

The ability of nitrogen fixing organisms, nitrifying bacteria and amonifiers of soils are seriously affected. Acid rains are also responsible for physiological malfunctioning and reduces certain species of lower and higher plants. The growth of certain vegetables such as radish, beans is also impaired.

Effects on Water Resources

The type of soil determines the effect of acid precipitation. Alkaline soils neutralize the acid deposition. Contrary to this, the sandy soils or granite terrain has less neutralizing capacity and aid deposition runs off into the rivers and lakes with most of its acid intact. A sudden release of huge amounts of acidic water into the rivers can be deadly dangerous particularly for fish. If it is acidic, all the original life in the rivers like snails, crustaceans and many insects will be destroyed. This will give rise to entirely a new ecosystem.

Toxic elements in a acidic runoff will contaminate both the surface and subsurface waters. Elements like aluminium, arsenic mercury, lead will be highly toxic to fish. The fish can die out by aluminium poisoning as soon as it hatches. Acid in water can also kill fish eggs. Salmon and trout fish species are very sensitive to acidification.

The critical pH level for most aquatic life is 6.0. When the pH falls below this level, a number of species of algae, insects, insect larvae and fish decline. Below pH 5.5 the number of snails and microscopic plant species are reduced. Below pH 5.2, snails disappear and the acidification affects the growth and productivity of aquatic plants, this is turn reduces the source of food for birds and animals.

Effects of Forest Resources

The close relationship between acid pollution and die-back of trees has been well established in Europe. More than 5 million hectares of forests have been severely influenced by acid damage alone or in combination with other pollutants in West Germany, Poland, Yogoslavia and Czechoslavkia. The conifers that grow in Germany are also affected by acid rains. In cases of severe acid damage, the symptoms include yellowing and loss of needless, stunted growth, drying out of tree tops, soft rot and root damage.

Damage to Animals and Wildlife

The wildlife and domesticated cattle population survives on a highly stable and complex food chain. The forest floor and flora forms the foundation of their food chain in the shape of herbaceous element. Once the forest floor and its flora is damaged, the whole food chain is broken that affects finally the stability of the animals and wildlife population at various trophic levels of the ecosystem.

Damage to Human Health

The ailments which are related to acid rains, are asthma, bronchitis, cancer of lungs and damage to the central nervous system.

Fluorosis in livestock is due to the excessive intake of fluoride through water. The contaminated milk of diseased animals when consumed by the human beings also causes fluorosis. The great concentration of mercury in water is responsible for thyroid abnormalities, liver disorders, paralysis, pulmonary oedema and corneal disorders.

Developmental Activities

Large area of fertile and productive crop lands, forest areas, grasslands and horticulture are encroached upon by various developmental activities such as rapid urbanisational and human settlements, mining/quarrying, construction of factories, industries, dams, canals, roads, railways, airports, etc. This thoughtless action of our government is jeoparadising the food production for future generations.

Control Measures

Control of soil pollution is largely connected with fertilizers and pesticides application, and solid waste disposal. Control measures to check soil pollution are :

(i) The manufacture and use of chemical fertilizers should be minimized and instead of them bio-fertilizers and well decomposed organic manures should be used.

(ii) Biological and ecological methods of pest control should be encouraged to reduce the use of pesticides.

(iii) Salinity/sodicity should be prevented by providing the flood prone, and irrigated lands with proper drainage.

The management of salt-affected soils can also be done by growing crops tolerant to salinity and sodicity like barley and oats; tree species (*Acacia nilotica, Acacia modesta*) and adopting better water management practices. Application of organic manures, including green manuring has been found very effective for reclamation of such soils. Use of Dhaincha (*Sesbania aculeata*) as green manuring has been found to be best methods for reclamation of sodic soils under R.S. Pura (Jammu) conditions (Najjar, 1993). This species in addition to serving as green manuring has also proved beneficial in providing fuel wood to the tune of 15 tonnes ha^{-1} on dry weight basis after about 150 days of its growth.

(iv) Control of land degradation can be done through restoring forest and grass cover as also to check erosion and floods. Reforestation *i.e.* growing of suitable tree species to give proper cover in the eroded areas is the simplest and easiest way for checking soil erosion. Planting of shrubs such as *Carissa spinarum, Adhatoda vasica, Desmodium* spp., *Mimosa himalayana* in eroded areas and community waste lands, is useful to stop erosion. These shrubs serve the dual purpose of meeting the villagers fire wood needs and helping soil/ water conservation. Grasses such as *Cenchrus* and a legume *Stylosanthes* should be given due place to grow in the eroded areas for supplementing the fodder and erosion control.

(v) Solid and liquid excreta of livestock can be used for preparing organic manures. But the common practice of storing manure on cultivator's field is very defective. Dung and litter are usually collected each morning and put in heaps in open spaces of the fields. The manure heap as such remains exposed to rains and sun shines, which as a result creates considerable loss of nutrients. Approximately half of the manurial value is lost. During decomposition, there is loss of ammonia by volatilization. The following scientific method is recommended for preparing good quality farmyard manure and to avoid high nutrient loss.

(1) The manure preparation should be carried out in trenches of 6 – 7.5 m long, 1.5 – 2 m broad and 1.0 m deep. All suitable dry litter and farm refuse should be kept near the cattleshed and about 2 kg (mixed with earth) should be spread under the feet of each animal in the evening for absorption of urine. The litter should be localised in the areas where urine generally drops and soaks into the ground. Each morning urine soaked litter and dung should be well mixed and taken to the manure trench. A section of 1m length of the trench is taken up from one end for filling with daily collection of refuse. When the section is filed 45-60 cm above ground level, the top surface is plastered with cow dung earth slurry, after which the next 1 m length of the trench is filled.

(2) When the trench is completely filled, the second trench is made and filled in a similar manner. In about 3-4 months, the manure is ready for use. Normally by the time the second trench is filled, the manure of the first trench would be ready to use. The number of trenches needed would, however, depend upon the number of animals. Generally, two such trenches would be required to 3 to 4 cattle. By following this method it is possible to prepare about 5-6 tonnes of manure per year per head of cattle.

(vi) Recycling the recovery of materials is another reasonable solution for reducing soil pollution. This would decrease the volume of the refuse and help in the conservation of natural resources. As for example, materials such as paper, grass and some kinds of plastic can be recycled. Although the cost of recycling of paper and glass is high, it is worthwhile in terms of resource conservation. It has been found that recovery of one tonne of paper by this method would atleast save 17 trees which otherwise would be cut and used for getting this much amount of paper.

Priorities and Suggestions for Future Research Work

(i) Cultivation of crops on steep slopes should be totally banned.

(ii) Use of chemical fertilizers should be restricted whereas that of organic manures (prepared on scientifically) must be encouraged.

(iii) The need of the hour is to conduct more research on bio-fertilizers (Rhizobium, azotobacter culture, blue green algae inoculant, azolla etc.) to cut short the need of chemical fertilizers for various crops.

(iv) Research is also needed for working out the various substitutes for insecticides and fungicides to be used for control of insects and fungal diseases in various crops. For this purpose, natural resources like Neem tree, Drek and others can be exploited.

(v) Hand weeding should be preferred over the use of weedicides for control of weeds.

(vi) Eroded areas call for immediate rehabilitation measures. These include afforestation and reforestation. The selected plant species must be free from attack of diseases and insects. They must possess high survival rate under adverse conditions of frost and drought. The selected trees must have efficient nutrient pump *i.e.* they should have ability to bring nutrients from lower horizons to the surface soil. They should be deep rooted.

(vii) Research efforts have to be intensified for delineating the problematic soils and evolving suitable management practices.

Impact Points to Remember for Soil Pollution

- Although soil can depollute organic matter in its various forms by decomposing it into harmless products like humic substances, carbon dioxide and water yet organic compounds like DDT, BHC and many others are resistant to decomposition. As such they are hazardous to human/animal health.
- The ability of soil to act as depollutant is due to the presence of organic and inorganic colloids in it.
- Organic manures when not prepared in pits but thrown as heaps in open fields, create air and water pollution also.
- Chemical pollution of the soil results from the unintended or incidental contamination of the soil with man made chemicals.
- Nitrate form of nitrogen (NO_3) is more mobile in soils, so excessive use of nitrogenous fertilizers in soils, causes its

accumulation in well waters and lakes. In lakes its accumulation alongwith phosphates produces eutrophication.

- When concentration of NO_3 in surface or underground drinking water exceeds 90 ppm, it causes diarrhoea in human beings. In children, it produces blue blood baby disease when its amounts in drinking water exceeds 45 ppm.
- The danger of NO_3 pollution is much more in sandy soils owing to high percolation capacity.
- Fodders like oats, maize stalks when raised on soils with heavy application of chemical can cause health problems to animals.
- Fluorosis in livestock is due to excessive intake of fluoride through water.
- Biological and cultural methods of pest control should be encouraged to check soil pollution.
- Biofertilizers and organic manures be used instead of chemical fertilizers.
- Land degradation can be checked by restoring forests and grasslands.
- Afforestation is the simplest and easiest way to check soil erosion.

REFERENCES

Gupta, R.D.; Marwaha, B.C. and Bali, S.V. (1985). Soils of Jammu and Kashmir and their management. In *Soils of India and their Management*, FAI, New Delhi, pp. 164-176.

Gupta, R.D.; Gupta, J.P. and Singh, H. (1992). Problems of Agriculture in the Kashmir Himalayas. In *Conserving Indian Environment* (Ed. By S.K. Chadha) Pointer Publisher, Jaipur, India, pp. 53-63.

Gupta, R.D. (2002). Checking soil pollution. *Daily Excellsior* 38(119):6.

Gupta, R.D.; Bali, S.V.; Singh, H.; Khajuria, M.R.; and Kaul, V.K. (1990). Ecology of Kandi belt of the outer Himalayas of Jammu division, Jammu and Kashmir State. In *Himalayan Environmental Problems* (Ed. by S.K. Chadha) Ashish Publishing House, 6/81, Punjabi Bagh, New Delhi.

Gupta, R.D. (2002). Save the soil, save the nation. *Environment and People*. 8(12): 20-25.

Jamwal, J. (2002). Wasteland development in Jammu Region. Personal Communication, Trikuta Nagar, Sector 1, Jammu.

Kanwar, J.S. (1999). Soil Science, food security, sustainability and the society. *J. Indian Soc. of Soil Science* 47(4): 634-648.

Najar, G.R. (1993). Characterisation and management of salt affected soils of Jammu district. M. Sc. Thesis Agri. (Soil Science), Faculty of Agri., SKUAST, Srinagar.

W.H.O. (1984). *Guidelines for drinking water quality*. Vol. 1. recommendations, World Health Organisation, Geneva, pp. 1-130.

9

PESTICIDAL POLLUTION

Introduction

The use of chemicals to kill pests is not a new concept but it is as old as 70 A.D., when Pliny Elder gobbled an idea for destroying the insects with arsenic compounds. The Chinese elaborated this notion sparingly to use arsenic sulphide as an insecticide during late sixteenth century. The use of arsenic compounds continued till during twentieth century. The use of synthetic organic pesticides, however, gained momentum around 1940 with the coming up of a large number of manufacturing and formulating units all over the world after the development of Dichloro Diphenyl Trichloroethane (DDT). Although D.D.T. was first discovered by O. Zeidler in 1874 yet P. Miller working for Geigy in Switzerland discovered that it a has remarkable insecticidal properties.

In agriculture, use of various pesticides was started with the introduction of high yielding varieties of different crops during late sixty of twentieth century. In fact, the new introduced varieties of wheat, rice, sugarcane, oilseeds, millets and many others possessed more response to applied nitrogenous, phosphatic and potassic fertilizers, and as a result they are liable to become more susceptible to an assault of different insects and diseases. Crop loss due to pest attack ranged from 10-30 per cent depending upon the crop and environment. Annual crop losses due to pests in India are estimated at Rs. 2,00,000 million per year (Saxena, 1996). Thus, the use of pesticide has now become an increasingly necessary operation for controlling pests and in the consistent and economic production of the crops throughout the world.

Pesticides Consumption

Now the pesticide problem associated with intensive agriculture is the most serious one in many countries and requires to be tackled with care. India being the largest producer as well as consumer of pesticides in the South Asia, tackling the problem of pesticidal pollution is all the more important. The total quantity of pesticides used in the country was merely 2350 tonnes during 1955-1956 which rose to 75,890 tonnes during 1988-1989, and 86,000 tonnes during 1990-1991 (Saxena, 1996). Subsequently its consumption declined gradually and during 1996-97, the total quantity of pesticides excluding herbicides consumed was only 5,611 tonnes (Suba Rao, 1999). At present, use of pesticides in India is hardly 0.03 to 0.56 kg ha^{-1} which is far less than the consumption of more than a kg ha^{-1} in U.S.A. and 10-12 ha^{-1} in Japan (Tomar and Parmar, 1993). Yet, the problem of pesticide use has assumed alarming proportions due to eratic use of pesticides in various regions, cropping systems, crops and their indiscriminate use, misuse and abuse.

In rainfed areas which constitute nearly 70 per cent of the total cropped area of the country, the pesticide use is the minimum, but in irrigated areas its use has been high (Table 9.1). Cotton, rice, fruits and vegetables which account for less than 35 per cent of the cropped area, consume 84 per cent of the pesticides. On the other hand, wheat, coarse cereals, millets and pulses, covering 54 per cent of the total cropped area consume only 3 per cent of the pesticides used. The data on spatial distribution of pesticide consumption in agriculture are shown in Table 9.2. It is worthwhile to mention that about 18.0 per cent of the total consumption of pesticides in India is concentrated only in Andhra Pradesh which though accounts for 7.0 per cent of the gross cultivated area (Subba Rao, 1999).

Although pesticides are considered as panacea to contain pests yet the injudicious use of pesticides over the years has resulted in a number of serious detrimental side effects. Moreover, they have upset natural ecological balance and ecosystem.

Pesticide residues in soil, water, environment and foods are now of serious environmental concern. In this chapter, pesticidal pollution in terms of its definition, hazards and factors responsible for causing and managing pesticidal pollution have been elucidated.

Table 9.1: Pesticides usage on different crops in India

Crop	Pesticide use (%)	Cropped area (%)
Cotton	54	5
Rice	17	24
Vegetables & fruits	13	3
Plantation crops	8	2
Sugarcane	3	2
Oil seeds	2	10
Others (wheat, coarse millets, millets and pulses)	3	54
Total	100	100

Source: Gupta, 1986.

Table 9.2: Pesticide consumption in major States of India in relation to gross cultivated area

State	Pesticide consumption (Percentage)	Gross cultivated area (Percentage)
Andhra Pradesh	17.9	7.0
Uttar Pradesh	13.2	13.8
Punjab	11.8	4.2
Haryana	8.3	3.2
Maharashtra	8.3	11.4
Gujarat	7.4	5.4
Karnataka	6.4	6.4

Source: Subba Rao, I.V. (1999). *J. Indian Soc. Soil Sci.* 47 (4) : 614.

Definition of Pesticides

Broadly speaking pesticides are known as physical, chemical and biological agents which can kill the harmful creatures termed as pests. Under the Federal Environmental Pesticides Control Act, the pesticides have been defined as "Any substance or mixture of substances intended for preventing, destroying, repelling or mitigating pests viz., insects, bacteria, fungi, algae, nematodes, rodents, weedicides and other forms of terrestrial and aquatic plants or animals".

According to the New Columbia Encyclopaedia "Pesticides are biological, physical or chemical agents used to kill plants or animals that are harmful to man, in practice pesticide is often applied only to chemical agents".

Definition of Pesticidal Pollution

Pesticidal pollution may be defined "as the pollution caused by the excessive and indiscriminate use of pesticides owing to their hazardous effects on human, animal and plant life, through contaminating air, water and soil".

Presently, presence of high and toxic amounts of pesticidal residues in foodgrains, food stuffs, fruits, vegetables milk and milk products, eggs, soils and water have been detected. Even human breast milk is found to contain pesticidal residue. Such alarming pesticidal pollution is also reported from Indonesia (Naik, 1996), Bangladesh (Abdul Kalam, 1996) and several other countries.

Classification of Pesticides

Pesticides or the biological agricultural chemicals can be classified as:

- Insecticides
- Rodenticides
- Molluscicides
- Piscicides
- Nematicides
- Acaricides
- Fungicides
- Bactericides
- Alagecides
- Herbicides

Of these pesticides, insecticides are the most widely used pesticides which include DDT, endrin, dieldrin, chlordane, aldrin, endosulphan, isobenzene, metaoxychlor, lindane, BHC. These are referred to organo-chlorines. Next kind of insecticides are carbamates as zectran, carbofuran, fanuron. These get accumulated in food chains at various levels to adversely affect nervous system, muscular coordination and other disorders. There is another group of substances called organophosphatics such as parathion, malathion, monocrotophos, Menazon. These insecticides also affect nerve cells. Muscular debility and dizziness, are the other effects of these insecticides.

Rodenticides are other kinds of pesticides. These are directed against rodents. Examples are warfarin, sodium fluoroacetate. Snails slugs and moluscus are killed by another kind of pesticides, called molluscicides. Copper sulphate is one of the example of these pesticides. This reduces snail population. Molluscicides seriously interfere with structure of aquatic vegetation.

Pesticides are used to control undesirable fish. Some of these pesticides are Toxaphene, Antimycin and TMF. Fungicides are used to control fungi which bring about a number of diseases in plants. For example Bavistin is used to check loose smut of wheat. Nematicides are the compounds used for controlling nematodes *e.g.* Dichloro dibromopropane (DDBP) and carbofuran. Some of the chemical compounds are also used for managing mites and are known as acaricides. Similarly for eliminating undesirable plants *i.e.* weeds, herbicides are used.

Sources for Causing Pesticidal Pollution

Pesticidal pollution is caused mainly through excessive and indiscriminate use of pesticides, waste water, organic wastes, aerial spray of the pesticides, environmental imbalance pesticide resistance and others.

Manufacturing Process

Manufacturing of pesticides is a direct source for polluting the environment, thereby, endangering the life on earth. Union carbide had given a big jolt to Indian while manufacturing the carbamates insecticide (Carbaryl), leading to Bhopal gas tragedy in 1984. Besides causing several deaths in and around the factory site, its indirect effects are still visible in the affected population.

Indiscriminate and Excessive Use of Insecticides

The indiscriminate use of insecticides is one of the largest source for polluting the air. The toxic chemicals used for the preservation of foods also cause air pollution. The regular spraying of the insecticides pollute not only the air but also the water.

Since the quality chemicals were costly, the farmers and orchardists started using inferior quality insecticides which instead of providing beneficial effects, did great harm to the crops, vegetables and fruits. Not only the soil was subjected to frequent treatment of chemicals which as a result became less productive or totally unproductive. In fact the frequent use of insecticides made soils deficient in certain minerals, nutrients and other ingredients. The reckless use of pesticides in Punjab and Haryana had given rise to brown plant hopper a very common pest of paddy crop.

Industrial Effluents

The industrial effluents include poisons, chlorine, lead, mercury and other chemicals, which are more toxic than municipal wastes and pose a danger to people, animals and plants. For example, an effluent called Paris Green (copper acetoarsenate) has caused ground water poisoned in sections of Kolkata from a pesticide industry through percolation of arsenic. In Ludhiana, a thriving cycle industries, manufacturers of basic alloys, gas burners, many of them using electroplating processes, release effluents with high concentrations of cyanides and other chemicals which after mixing with water make it unfit for drinking. Similarly in Mirzapur, the carpet industry is responsible for release of toxic dye effluents which in turn have changed colour of well waters adjoining the dye industries.

Industrial Wastes

A number of organic and inorganic wastes produced by innumerable industries in megacities, are the other potent sources for causing pesticidal pollution. For example, Hindustan Insecticides Limited and Shri Ram Food and Fertilizers are two of the major industries which contribute the poisonous chemicals into the river Yamuna through wastes dumped in the nearby of this river. Other pesticides such as DDT escapes into rivers with factory wastes from industrial processes like moth proofing of wool. This can also happen from factory that produces DDT.

The water of river Ganga that takes its birth at Gangotri in Garhwal region of Uttaranchal Himalaya was thought to be pure and unpolluted up to Haridwar but now it is not the case. Even at Haridwar it is being polluted by the organic wastes of a number of industrial units. At Farukhabad water of the river Ganga becomes red due to discharge of industrial wastes. These wastes have toxic chemicals. At Kanpur, the river is worst affected by organic and inorganic chemicals from 55 tanneries, 15 textile mills, 2 jute mills and a number of other drug and chemical factories, which are discharged into it.

Animal Excreta/Slaughter House Wastes

Huge amounts of animal excreta accumulated in cattle sheds, piggeries and poultry farms and wastes from slaughter houses, contaminated with a number of pesticides, are discharged into grazing fields or dumped into pits. These are carried through run offs and surface leaching, particularly during the rainy season to cause water pollution.

Underground water which was once considered a sage source of water is now becoming increasingly polluted. This is primarily because disposal of effluents having pesticides, fertilizers and other chemical pollutants like compounds of arsenic, mercury and lead. Some compounds are so corrosive that they may even affect waste treatment plants.

Aerial Spraying

Pollution of air by pesticides is a man made problem. Pesticides reach the air from aerial spraying, when it is aimed at certain crops and sprayed in the air, it drifts according to direction and velocity of the wind into other areas. Handique (1987), has reported on the drift of chlorinated hydrocarbon spray applied from aircraft and on factors governing the amount of contamination to be expected on nearby forage crops. It was found that one kg of chlorinated pesticide under turbulent windy condition resulted in residues of 20 ppm in the fields of 73 m distance. Therefore, applying aerial sprays of bufferzone of adjacent fields must be kept and in case of the wind current, its velocity must also be considered.

Mines/Quarrying

Out of an estimated number of about 70,000 chemicals with an annual growth rate of 3,000, as many as 10,000 are believed to cause an irreversible damage to the biological world. Most of the harmful chemicals are released from various working processes of mines. The carbon tetrachloride and chloroform released from mineral plants can cause liver cancer.

The conventional methods of mining are actually responsible for causing environmental pollution. In the highly developed countries the mining operations are conducted strictly on the latest methods of mining technology and as such these methods are eco-friendly. However in the developing countries like India less scientific technology had been used and hence hundred of miners are buried alive in mines due to negligence mining activity.

The conditions are worst in Bihar and Bengal coal mines. There are heaps of wastes which are continuously damaging the fragile ecosystem of the region. The granite mining in south India, Bauxite mining in Bihar, Lignite in Assam and Jammu and Kashmir, quarrying in Ramban (J&K State), Dharamsala and Shimla (HP) and diamond mining in MP, have changed the environmental scenario of the country by way of producing pollution in air, water and soil.

Soils

It is well known that soil is the ultimate reservoir of pesticides. All chemicals when applied as soil application or when applied on plants herbs or weeds as aerial spray, reach the soil. Once they reach the soil, these are acted upon by a variety of physical and chemical factors, and microbial organisms.

Many chemicals such as hydrocarbon group of compounds *e.g.* DDT and BHC are non-biodegradable or if so biodegradation takes place very slowly. As such, their effect remains in the soils for a long period and, thus, DDT is a very persistent insecticide. Successive applications of DDT in the soil leads to its contamination to a great extent. DDT is, therefore, used in restricted manner in many parts of the world. Now in many countries, its use has totally banned because of its biological magnification. Nevertheless it is still used indiscriminately in many parts of India. In Assam, encephalitis has become a dreaded disease. In 1987, in Jorhat district 600 and in Sibsagar district 400 human lives were lost due to encephalitis. DDT has been used as quick method to kill the mosquitoes. The result is the alarming increase of DDT resistant strains of mosquitoes.

Pesticides reach the aquatic environment either through direct application to control aquatic pest or through erosion resulting from heavy rains or from irrigation. Pesticides are carried with the soil particles and get deposited in the bed or streams, ponds and on land areas which are subjected to flooding. The rate of movement of pesticides depends on soil type and pesticides, water solubility and its chemical and physical nature.

Hazards of Pesticidal Pollution

Hitherto pesticides were, as referred to be used in small quantities, thus without releasing any adverse and toxic effects on the ecosystems. The excessive use of pesticides has, however, now become a serious matter because of their health hazards to human beings, animals and plants *vis-à-vis* their effects on wildlife. Pesticidal pollution is thus becoming a major environmental pollution and its abatement must receive attention to all.

The adverse effects can be categorized into:

(i) First order impacts involving pollution of the air by spraying of chemicals such as oxides of sulphur and nitrogen. The spraying of pesticides also results in fluctuating the atmospheric temperature, pressure and global warming.

(ii) Second order impacts involving contamination of soils, choking of surface drainage systems and spread of pulmonary diseases among residents around the factories and industries.

(iii) Higher order impacts resulting from dispersion of large amounts of heavy metals released during combustion and out of solid waste disposal.

The toxicological effects can cause serious metabolic and physiological disorders. In certain cases reproductive potential is lost and some compounds cause neurological disorders, behavioural disorders, paralysis. Some of the pesticides and fungicides have effects on digestive system leading to hepatitis disorders and cancerous problems.

The direct hazards are occupational, and they affect the farmers and labourers directly using them. As a matter of fact, those who use the chemicals are mostly illiterate and they do not know the right use of the pesticides. The farmers even sometimes do not wash their hands, do not use masks while spraying pesticides, do not have adequate quantity of water in the pesticides and also do not understand the right time of using these chemicals in their fields. As a result of these ignorances the insecticides do more harm than good to the crops.

Children and aged persons are quickly attacked and subjected to diverse types of diseases like skin troubles, breathlessness, hypertension, high blood pressure, baldness, eye disorders, and scabies.

Some of the agro-chemicals leave behind residues in the food, thereby, producing ill effects when the concentrations exceed the safe tolerance levels. Different food commodities are found to contain high levels of pesticide residues of DDT, BHC and sometimes even lindane, dieldrin and malathion in them. Indeed, DDT and BHC do not decompose readily and remain contaminated with the fruits, vegetables, grains and fodders. On being consumed by livestock and human beings they enter the food chain and damage liver, kidney and other parts of human and livestock. The pregnant women when affected by DDT and BHC and other insecticides deliver deformed babies or still children.

Pesticides like parathion, malathion, menazonorganophates affect nerve cell, muscular debility, dizziness and a number of other disorders. Carbamates such as zectran, fenuron and carbofuran get accumulated in food chains at various levels to adversely affect nervous system, muscular coordination and other disorders.

An average Indian's body has the highest DDT level in the world varying from 12.8 to 3.1 ppm. Cases of aluminium phosphide poisoning deaths and incidents of poisoning due to other pesticides are reported from several parts of the country off and on. In an estimate, every year about 10,000 people die and about 40,00,000 suffer from various effects of pesticides. Pesticides have a tendency to persist and accumulate in the soil, affecting various food chains. The pesticides affect soil fertility negatively and causes toxicity. They damage the soil structure and sub-soil ecology. Their metabolites can even migrate along the soil profile in vertical as well as horizontal directions. It is mostly organic and inorganic matter of the soil that can adsorb and retain the pesticides.

The livestock provide highly infected milk which does more harm than good. Besides, they also prove toxic to many useful birds and insects like honey bees. The misuse of the insecticides accumulate along the blood vessels and put hurdles in the free flow of blood thus causing heart trouble and cardiac disorders. The fodder or food crops when sprayed with chemicals deliver poisonous effect in milk.

Application of various pesticides to crops, fruits, vegetables, forests or any habitat may not only result in extermination of pest species but also the non-target species like birds inhabiting the area. Bakhre (1987) have reported residues of lindane, BHC and DDT in body organs of few species of Indian birds like pigeon, crow, kite, vulture and cattle. In 1981, sixty Canada geese, six snow geese and many others were found dead on the south shore of Lake Texas due to spraying of parathion in wheat fields. In England, the birds having the highest concentration of organochlorine residues were fish eating birds.

Management of Pesticidal Pollution

Practically it is not possible to get rid of all the pests which live on the plants because natural balance will be upset if all the insects are destroyed. In the biological cycle, all insects depend upon each other and if a particular stock of insects is perished, the others will be deprived of the food. Every category of species have to exist for maintenance of ecological balance. In other words, each pest being controlled by some other creature for which it serves as food. Although an adequate dose of pesticides is used to control various pests to increase the yield of crops, vegetables and fruits yet over the years, the injudicious use of pesticides has resulted in a number of serious detrimental effects. Previously, there were minor hints of trouble but the use of organochlorines and organophosphates has added new dimensions to pesticidal pollution. Stress is being laid now on integrated control, by using different methods or techniques. These include cultural, physical, chemical and biological.

Cultural Control

It is one of the ancient and cheapest modes of pest control. Traditionally, farmers with their vast experience used to manage most of pest problems using these methods which consist of sanitation, tillage/removal of weeds, adjusting sowing/harvesting time, crop rotations, trap crops, water and fertilizer management, and host plant resistance.

Sanitation

It involves destruction of habitats where pests breed or hibernate. Destruction of crop refuse, disposal of garbage, fallen twigs and fruits from orchards, removal of diseased plants and pruning of infested plant parts offer an economic and effective methods to check crop pests and diseases. For instance, hibernating pink boll worm larvae in double seeds of cotton are easily killed if cotton seeds are crushed or fumigated in ginneries. Large number of over wintering sugarcane borers may be killed by ploughing under shaved stubs and crop residues. Clean field borders reduce the chances of building up of the pollution of pests like sorghum grey weevils and midge.

Tillage/Removal of Weeds

Ploughing and tillage expose soil inhabiting insects and destroy volunteer plants and weeds which provide food and breeding sites for many pests. Summer ploughing results in high mortaiity of cutworms, white grubs, grasshopper and many other soil inhabiting caterpillars and beetles. A number of insects are controlled by soil tillage, provided the life history and behaviour of the pest is known. Keeping the fields fallow in the season results in starvation and death of pests, thus, causing less infestation in the following year.

Kinds of implements used in tillage operation are also important in checking insects. Mould board ploughs are proved more useful in killing hibernating insects and exposing eggs and larvae to the surface than *desi* ploughs. Frequent harrowing hoeing and weeding also assist in destroying insects.

Adjusting Sowing/Harvesting Time

By adjusting the time of planting and sowing, infestation of some pests can be avoided. For example, early sowing of chickpea or mustard in the season can shun the attack of pod borer and aphid in respective crops. Early sown pea avoids pod borer damage as the pods are ready when temperature is low.

Crop Rotation

Crop rotation is one of the important technique to reduce pest problems. For example, rotating wheat crop after berseem reduces the attack of *Phalaris minor* in the following year. Rotating groundnut with maize or small millet reduces the attack of white grubs as these crops are poor food for the grubs, and adults lay fewer eggs on these crops.

Physical and Mechanical Control

The direct or indirect non-chemical methods used to destroy pests out rightly or to make the environment unacceptable to them are grouped under this category. They are the oldest and some of them are primitive, and hence often considered outdated by many people. However, under specific conditions they are still practised. There are some very successful examples for the use of these methods. Hot air treatment of sugarcane planting material is used successfully to kill inoculum of many diseases.

Soil solarisation is practised largely to kill soil borne pathogens and nematodes before planting betel vines. Cold storage of potatoes takes care of potato tuber moth population. Use of light traps is used for attracting white grub beetles in large number in hot spot areas.

Covering pomegranate fruits for protection against butterfly attack, greese banding mango trunks to prevent mealy bugs and use of non-chemical adhesive traps for rodent control in godowns and houses are leading methods.

Pheromones and Growth Regulators

The pheromones are mostly used as tools for monitoring natural populations but in some cases they are also effective in controlling pests. Pink bolloworm of cotton is one such example which has been effectively controlled in India using pheromone called "gossyplure", by confusing wild males.

Growth and development of insects have regulated by hormones. Some of them have been synthesized and used in pest management. These include juvenile hormone and ecdysone and chitin inhibitors. They interrupt normal growth process and lead to the development of abnormal adults which are unable to reproduce. They are environmentally safe being non-toxic to human and other higher animals. Rotating pigeonpea or chickpea with non-leguminous crops is a good example for managing Fusarium wilt and nematode problem.

Trap Crops

Use of trap crops is a good method for diverting the attention of the pest from the main crop. Mustard which is preferred by aphids, diamond back moth and cabbage leaf weber is successfully used as trap crop in the management of cabbage pests. These pests are attracted to mustard and cabbage grows almost pest free without any other treatment. Similarly castor is used as a trap crop for attracting tobacco caterpillar and white flies in tobacco field. Planting trap rows of tobacco or marigold in cotton field is used for diverting the egg laying females of *Helicoverpa* on these crops.

Soil Water and Fertilizer Management

Soil rich in humus with balance status of macro (NPK) and micronutrients (Fe, Cu, Zn, Mn, B, Mo and Cl) can withstand pest attacks due to its increased compensatory power. On the other hand soils rich in N but deficient in P and K can lead to more attack of pests like aphids, jassids and whiteflies. Similarly, appropriate soil moisture is important for maintaining good plant health and declining pest attack.

Host Plant Resistance

Use of resistant plant varieties of various crops is one of the most effective, economical and environmental friendly way. This characteristics is generally owing to some biochemical or morphological properties of plants, influencing behaviour or metabolism of pests causing thereby less damage to plants. As this method is compatible with other desirable ecofriendly methods of pest suppression like biological and cultural methods, host plant resistance is of special significance in Integrated Pest Management (IPM).

Botanical Pesticides

Plant based insecticides like *neem* offer good promise in pest management as they are mild, naturally occurring chemicals with no or very little adverse environmental impact. In America, where a lot of awareness exists about devastating consequences of pesticidal pollution, *neem* extractions of different strengths and blended alongwith other milder and degradable chemicals are used to maintain crops.

In India too, many *neem* based formulations are available. Even freshly prepared *neem* seed kernel extract is also utilized for controlling different harmful pests. The development of low cost technology for using *neem* is very essential for the resource poor farmers. Farmers can easily prepare *neem* seed powder, *neem* oil and aqueous *neem* seed extract themselves. A number of commercial formulations are now available in the country (Table 9.3). Apart from *neem* several other plant species like pongomia, sweet flag, melia and garlic have shown promise.

Biological Control

In nature, pests are kept under control largely by their natural enemies and a equililbrium is maintained over a period of time where neither the pest nor their enemies increase beyond a certain level. However, if this delicate ecological balance is upset either by introducing more susceptible plant types or indiscriminate killing of natural enemies, the pest population increases. Under such a disturbed ecosystem, the pests can be regulated by using biotic agents and the process is known as *biological control.*

Biological control being relatively more permanent, safe and economical, possesses a number of advantages than other methods of pest control. Hence, lot of emphasis has been given in India to biological control of insects, diseases and weed pests. There are three kinds of approaches in biological control. These include classical biological control, conservation of natural enemies and mass rearing release.

Table 9.3 : Neem based insecticides

Name	Crops in which used	Pests controlled	Dose Litres (ha)
Margocide	Cotton, tomato & cabbage	Jassid, white fly & Aphids	5.0
-do-	Cotton	Boll worms	2.5
-do-	Tomato	Fruit borer	2.5
Neembecidine	Paddy	Gallmidge	0.4-0.5
-do-	Cotton	Aphids Leaf roller	0.4-0.7
Bioneem	Cotton	Aphid, Jassid, Bollworm,	2.5-3.0
Neembasol	Cotton	White fly, Bolworm, Leaf roller	1.5
Neemazal F	Cotton	Aphid, hopper, White fly	1.0
Neem Gold	Tobacco	*Spodoptera litura*	1.25
-do-	Coffee	Stem borer	1.25

Source: Saxena, R.D. (1996). Report on awareness programme on pesticides and sustainable agriculture. 228, Sidharth Enclave, New Delhi.

Classical Biological Control

In this method the natural enemies of an introduced pest are searched in their native home and released in the new area. Most of the spectacular successes achieved in biological control are in the field of classical biologicai control. In India, success achieved in controlling water hyacinth, water fern, *subabul* psylla by means of ntural enemies are some of the examples of classical biological control.

Conservation of Natural Enemies

The conservation of natural enemies is the most important step in biological control as if they are conserved, the need for other control measures is mostly reduced.

Conservation refers to the avoidance of measures that destroy natural enemies and promote measures which are suitable for their reproduction and development. To shun indiscriminate use of broad spectrum pesticides over large areas, to adjust spray schedules for causing minimum harm to natural enemies and to encourage natural enemy activity growing plants for alluring natural enemies, are some of the methods of conservation.

One of the most useful examples of this method is collection and preservation of cocoons of *Epiricania melanoluca* which is a parasite of sugarcane pyrilla at the end of the season and augmentive releases at the beginning of next season. Sowing of coriander or marigold in vegetable fields increases the activity of syrphid flies which predate on aphids and other soft body insects, thereby, reducing the pest numbers on the main crop.

Mass Rearing and Release

Mass production of many useful parasites and predators under controlled conditions and thereafter, releasing them into the fields to control pests, is termed as mass rearing and release technique. In India successful examples where mass rearing techniques for natural enemies have been developed, are use of different species of parasitic wasp, *Trichogramma* used to control caterpillar pests of sugarcane, cotton, paddy and tobacco. Antagonistic fungus,

Trichoderma harzianum is used as a wide spectrum bio-agent for control of many seed and soil borne plant diseases. The mass production of *Trichoderma* is done on moistened crushed grains of sorghum, maize or rice mixed with one per cent yeast extract after sterilizing in autocleavable in polythene bags.

Microbial Control

Use of pathogenic micro-organisms like viruses, bacteria, fungi, nematods and others for the control of crop pests and diseases is known as microbial control. They do not leave undesirable residues and can be used even close to harvest. Microbial control is also compatible with other methods of pest management. Some of the microbial agents which have shown potential and offer an excellent alternative to chemical pesticides are: *Bacillus thuringiensis, Bacillus popillae, Nuclear Polyhidrosis viruses* (NPV) and *Granulosis virus* (GV).

Bacillus thuringiensis is the most commonly used microbial pesticide for control of a wide range of pests having attack on cotton, pulses, tobacco and vegetable crops. Many commercial formulations are now available in our country.

Bacillus popillae is a bacterium effective against white grubs which are destructive polyphagous pests damaging groundnut, sugarcane and sorghum.

In India NPVs for *Helicoverpa armigera* commonly known as Ha NPV and that of *Spodoptera litura* called Sl NPV are available. The latter is recommended for the control of *Spodoptera* on tobacco and cotton and Ha NPV is recommended against *Helicoverpa armigera* chickpea, pigeon pea and cotton.

The GV for sugarcane tissue borer has been isolated and is available for field use.

Suggestions for Further Improvement

- Summer ploughing, destruction of crop residue and practicing clean cultivation should be done in order to destroy hibernating and soil inhabitating insects, pathogens and nematodes.
- Selection of healthy seeds of resistant varieties and preventive seed treatment for destroying initial disease inoculation is the

need of the day in order to minimize pest load at later stage.

- Proper water and fertilizer management and providing wider spacing for increased plant vigour should be followed to withstand pest attack and allowing better multiplication of natural enemies.
- Inter cropping and crop rotations should be strictly followed for promoting natural enemies and reducing pest population.
- Integrated pest management programme should be popularized among the farming community for its fully adoption.
- Need based judicious use of chemical pesticides is one of the vital aspects of crop pest management these days which must be taken with extreme care to minimise disturbance of these ecosystem.
- Use of botanical pesticides should be encouraged especially of *neem* based products of different formulations.
- Suppression of weeds through biological agents is one of the important and high priority these days, so it should be encouraged.

Impact Points to Remember for Pesticidal Pollution

- The term pesticides includes insecticides, fungicides, bactericides, herbicides, nematicides, rodenticides and soil fumigants. They cover a wide range of formulations: Chlorinated hydrocarbons, organophosphates, metallic salts, acetic acid derivatives, thiocarbamates.
- Many pesticides are non-degradable and their residues have a long life. The most persistent ones are organochlorine compounds which pass through food chains and mainly accumulate in fatty tissues and affect the nervous system.
- An estimate made by the International Labour Organisation (ILO) has indicated that pesticides sales have peaked worldwide at 27 billion US dollars. Industrialised countries use 80 per cent of world's agrochemicals but probably suffer only one per cent deaths from pesticide poisoning.
- Developing countries, on the contrary, suffer from 99 per cent of all such deaths, while using 20 per cent of the world's agrochemicals.
- The disparity stems not only from climatic conditions but more importantly from general poverty, malnutrition, ignorance, lack

of adequate safety measures and widespread lack of awareness of the hazards involved in handling of pesticides.

- India is one of the few countries still manufacturing and using the most toxic chlorinated pesticides which arc targetted for global phase out. At present, in India about 86,000 tonnes of pesticides are used and about 5 million women/farm workers are exposed to the dangers pesticides.
- Indians have the highest amount of pesticide residues in their bodies as compared to other population groups in the world.
- Indiscriminate use of hazardous pesticides such as DDT and BHC in farms and for preventive public health has resulted in an intolerably high level of pesticide residues in grains, fruits, vegetables and milk.
- According to neurologists of Lucknow based King George Medical College, these residues get stored in the tiny fat globules in the body of humans and animals. In the long run these affect nervous functions leading to impaired weakness and tremors.
- Indiscriminate use of pesticides has adversely affected the ecological balance resulting in pest resurgence, pesticide resistance, pollution of environment and food chain.
- The vastly stepped up application of synthetic pesticide has been accompanied by the massive resurgence of brown plant hopper the worst enemy of paddy fields. In 1977, Indonesia lost nine million tonnes of rice due to the brown plant hopper attack. Thailand has the largest hopper outbreak in 1989-90.
- In India, the reckless use of pesticides has given rise to brown hopper, a very common pest of paddy crop. This was more so in Punjab and Haryana.
- Besides playing havoc with the rise field eco-system, pesticides can adversely affect the health of rice growing farmers. In the Philippines, of the 4,031 acute poisoning cases reported by Hospitals between 1980 and 1987, 603 proved fatal. So a number of pesticides including endosulfan were banned by Philippines.
- A recent detailed study of pesticides and immune system by the Washington based World Resource Institute (WRI) identified hundreds of cases which showed that many widely used

insecticides, herbicides and fungicides can suppress normal immune system response.

- In USA convincing evidence exists that a variety of chemicals released by insecticides, have found in soil, well water, oceans, fresh water bodies and rain water.
- According to scientists of Haryana Agricultural University (India) 70 per cent of the cereals, water bodies, milk and milk products were contaminated with pesticide residues, of which 30-40 per cent contained toxic residues above the prescribed limits.
- They further studied that although the consumers could save themselves from hazards of ognanophosphates and carbamate pesticides in vegetables and fruits by applying simple processes like washing, peeling and cooking, yet contamination of milk with pesticides like BHC and DDT is difficult to reduce.
- The residue to BHC and DDT, being fat soluble get concentrated in butter and *ghee* and their amount usually goes beyond the prescribed safe limit. So the use of these insecticides as well as aldrin, chlordane and toxaphene has been banned in the country.
- Other pesticides found most harmful to ecology, are Dibromo-Chloro-Propane (DBCP), endrin, heptachlor, ethyl parathion, pentachlorophenol (PCP), tetradifen, phenyl mercury acetate, sodium cyanide, methyl bromide and ethylene dibromide which were banned by the Central Insecticides Board during 1993.
- Pesticides of chemical groups like organophosphates, carbamates and synthetic pyrethroids, however, being relatively easily degradable do not pose serious threat of widespread contamination on accumulation in the environment.
- Development of alternative farming technology encompassing use of low-cost, non-hazardous, locally available material like cowdung, *neem* and turmeric is the imperative need of the hour.
- Use of biopesticides, pesticides of plant origin as stated above, biocontrol agents and biotechnology would definitely help to curtail the use of synthetic pesticides in future. Integrated Pest Management programme must be strengthened.

REFERENCES

Abdul Kalam, K.M.H. (1996). Report of an Asian Productivity. Organization study meeting 25 June – 5 July, 1996, Tokyo, Japan, pp. 141.

Bakhrc, P. (1987). Pesticidal Polltuion of the Environment and its impact on birds. In environment and pesticide toxicity. Proceedings of 8th Annual Session Acad. *Environment Biology and Symposium* (Ed. by Dalela *et al.*) Jammu University J&K, India.

Gutpa, D.K. (1986). *Pesticides in Indian Environment.* Inter Print Publisher, New Delhi, ICMR, 1993. Indian Council of Medical Research, New Delhi – India.

Handique, R. (1987). Pesticides and the environment. In environment and pesticide toxicity. Proceedings of 8th Annual session Acad. *Environment Biology and Symposium* (Ed by Dalela *et al.*,) Jammu University, J&K, India.

Naik, S. (1996). Report of an Asian Productivity Organisation Study Meeting 25 June – 5 July, 1996. Tokyo, Japan, pp. 225.

Saxena, R.D. (1996). Report on awareness Programme on pesticides and sustainable agriculture. 228, *Sidharth Enclave*, New Delhi-110 014.

Subba Rao, I.V. (1999). Soil and Environment Pollution – A threat to sustainable agri. *J. Indian Soc. Soil Sci.* 47(4) : 611-633.

Tomar, S.S. and Parmar, (1993). Evolution of agrochemicals in India - An overview. Indian Agricultural Research Institute, New Delhi, India.

10

RADIOACTIVE POLLUTION

Introduction

Radioactive pollution, pertaining to all major life supporting system *viz.*, air, water and soil, is a special form of physical pollution. It is imperative to elaborate radioactive pollution separately because its nature of contamination is quite different from those of others. Its effects are also of special kinds. Radioactivity is a phenomenon of spontaneous emission of protons (alpha particles), electrons (beta particles) and gamma rays (electromagnetic waves having short wavelength) as a result of disintegration of atomic nuclei of certain elements.

Sources of Radioactive Pollution

People have been exposed to low levels of radiation from natural sources for several millennia but its danger has increased recently. Sources of environmental radiation are both natural and artificial (man made). The former consists of cosmic rays which reach the surface of the earth from space and terrestrial radiations from radio-nuclides present in the earth's crust. Many radio-nuclides such as radium-224, uranium-235, uranium-238, thorium-232, radon-222, potassium-40 and carbon-14 occur naturally in rocks, soils and waters.

Man made sources of radiation pollution are mining and refining of plutonium and thorium production, and explosion of nuclear weapons, nuclear power plants and fuels, and preparation of radioactive isotopes.

Nuclear Weapons

Nuclear radiations from various sources constantly pose an environmental risk. The first atomic bomb was exploded in Hiroshima followed by the second in Nagasaki (Japan) during 1945 on 6th and 9th August, respectively. These heinous events wiped out large number of innocent people *vis-à-vis* animals and plants. Despite the great suffering during 1945, the nuclear race by the big powers is still continuing unabated. The same happened due to the use of nuclear bombs in Vietnam, Gulf countries, Kuwait, Iraq. Presently, the stock file has enough energy to completely destroy the whole universe.

Tests of nuclear arms comprise the use of uranium-235 and plutonium-239 for fission while hydrogen and lithium are used as fission material. These tests give rise to very large neurton flux conditions which, in turn, render other surrounding materials radioactive. These materials compose mainly of strontium-90, cesium-137, iodine-131, and unused explosive and activated products. The radioactive materials are converted into gases and fine particles which eventually enter into the air like a mushroom cloud. These materials are too carried away by the wind and spread to vast areas, and on settling they cause water and soil pollution. When rain drops containing radioactive particles fall on the ground radioactivity is transferred to soil particles. From soils the radioactive materials enter the food chain influencing various life forms including human beings. In fact, water bodies after receiving radioactive substances through soluble products, transfer them to aquatic organisms and through food chain these ultimately reach man.

With the increased number of nuclear reactors all over the world reaching a figure of 420, there is always a lurking fear of the type of mishap which occurred at Chernobyl in Russia on 26th April, 1986. This incident not only devastated a vast area around Chernobyl which is still not safe for habitation, but also gave rise to a nuclear cloud which remained over a large part of Europe for several days. Even though only two to six per cent of the radioactivity contained in the core of the unit-4-reactor was released during the incident, this was enough to cause nearly half a million cancer cases. Besides, a large number of animals and plants were threatened to vanish in many parts of Europe. High levels of radioactivity continue to be detected in vegetation, milk, oil, food, water, air and soil.

From the foregoing paras, one can imagine the main menace to global environment by nuclear war. Those who have been the much talked movie "The Day After" will realize the horror posed by this scenario. It is said that when the story writer, Ed Hume, saw the movie, it sent a shiver down his spine and his thirteen year son turned pale.

Whenever, a nuclear incident, occurs dangerous radiations are emitted. Similarly, the debris, that are left in the atmosphere, after nuclear explosion, consist of dangerous particles which are harmful. For example, an isotope of strontium (Strontium-90) causes blood cancer and decays very slowly. Cesium-137 binds itself to soil and causes genetic change and iodine-131 has a tendency to cause thyroid cancer.

Radioactive Wastes

Wastes from atomic reactors containing various types of radioactive isotopes are most dangerous substances to deal with because their radioactivity cannot be destroyed at man's will. These wastes pose grave public health hazard wherever they are dumped. These wastes affect the aquatic plants and animals to a very great extent. They cause ionization of various body fluids, gene mutations and chromosomal mutations. They may accumulate in the bodies of aquatic animals like fish, thus harming fish as well as the animals and men which eat them.

Inert gases and halogens escape as vapours and become potential pollutants of the environment as they settle on land or are washed into surface water.

Radioactive Fallout

The radioactive dust that falls to the earth after atomic explosions is known as *radioactive fallout*. The kind of radioactive fallout varies according to the bomb exploded. Some of the radioactive elements like iodine-131 and phosphorus-32, have very short half life period. As such they decay fastly and cannot get into the metabolic pathways in plants. However, some radioactive elements have very long half life period *i.e.*, rate of decay is very low. For example calcium-45 has half life period of 160 days whereas strontium-90 and cesium-137 have still longer half life periods. Accordingly these substances with longer half life periods are potentially most dangerous isotopes to man and other vertebrates.

These radioactive elements enter into biological system by absorption through the leaves of the plants or through the roots of the plants. From the plants, they enter into the herbivores and from there to the cornivores upto the highest level of consumers in the food chain.

In terrestrial animals, strontium-90 is acquired chiefly through grazing food chain particularly in areas where the rainfall is high and concentration of other soil minerals are low. It may route to man via the grass-cow milk pathway or through the ingested plants and vegetables in which strontium-90 is concentrated.

Other Sources

A large number of radioactive isotopes such as carbon-14, phosphorus-32 and iodine-125 are widely used in scientific research. Waste water containing these radioactive materials reach the rivers and lakes through the sewers. Some of these isotopes such as radioactive iodine and phosphorus concentrate in slimes, sludges and microorganisms which enter the human food chain through fish and other aquatic life.

Exposures of Man to Radiations

Man is exposed to ionizing radiations from the following sources:

Industrial Applications

Luminous dials used for watches and other instruments, sterilization of foodstuff and drugs, radiographs of metallic articles, X-ray fluoroscopy, X-ray crystallography and spectroscopic analyses, are some of the kinds of this source.

Medical Applications

This source of radiation consists of tracers, internal therapeutic agents and external therapeutic techniques used in the body to examine the behaviour of some elements, to determine selective absorption of certain elements by specific tissues and to detect localisation of tumours/various diseases, respectively.

Civil Exposure

Man is exposed to slight amount of irradiation from cosmic rays, nuclear reactor accidents, X-rays generated by television sets and fallout from nuclear weapon testing which already stand discussed.

Mode of Action of Radiations

After intrusion of radiations in the human body, their primary effect is the production of ions followed by alteration in the chemical structure of enzymatic systems. Cells undergoing mitosis are more sensitive to damage. Chromosomes may also be damaged which finally induce malignancy and genetic abnormalities. The foetus is highly susceptible to radiation.

As already stated that risk is greater if the half life period of the radioactive element is long and the risk is less with those elements, which have short half life period and are more quickly eliminated. Briefly, the sources of radiation are shown in Table 10.1.

Effects of Radioactive Pollution

The effects of radioactive pollutants depend upon (i) energy releasing capacity, (ii) the halflife of the radioactive elements, (iii) rate of diffusion, and (iv) rate of deposition of the contaminant. Other factors influencing their effects are atmospheric and climatic elements such as temperature, rainfall and wind.

Almost all organisms are affected by radioactive pollution. Some species preferentially accumulate specific nuclides. For instance oysters can accumulate zinc-65, fish iron-55 and marine animals selectively take up strontium-90. Even milk and its various products can become highly polluted.

In the Northern Hemisphere, where most of the atomic explosions have occurred, strontium-90 has reached the highest levels (Kotpal and Bali 1989) which many biologists think are dangerously high. Strontium-90 is of greatest biological interest as it behaves like calcium and it incorporates quickly into biogeochemical cycles. In vertebrates, it becomes concentrated in tissue, where it can damage both to bones and bone marrow cells. It can also cause tumour.

Table 10.1: Sources of radioactive pollution

Natural	Man made
1. Cosmic rays	1. Medical and dental X-rays, Radioisotopes
2. Environmental	2. Occupational – Exposure
(a) Terrestrial	3. Nuclear – Radioactive fallout
(b) Atmosphere	4. Miscellaneous
	(a) Television sets
	(b) Radioactive dial watches, isotopes tagged products
	(c) Luminous markers
3. Internal	
(a) Potassium – 40	
(b) Carbon – 14	

Source: Park – 1997.

A number of studies have been conducted to investigate the radiation effects on human beings and other organisms. Radiations in higher doses (2000 rads or more) can cause death in few minutes to a few hours due to damage to the brain and nervous system. With less severe but lethal dose nausea vomiting appear immediately followed by diarrhoea. With sub-lethal dose acute symptoms are mild but after an interval leucopaenia (reduction in number of leucocytes in blood) and depression of all the cellular elements of the blood, diarrhoea and epilation appear.

Delayed effects have also been observed. These effects are seen many months or years later. These are :

(i) Malignant tumours and leukemia.
(ii) Skin burns, dermatitis, ulceration and infection.
(iii) Gangrene disease.
(iv) Lymphatic system: Leucopaenia.
(v) Bone marrow: All elements depressed leading to

granulocytosis, aplastic anaemia and thromocytopaenia.

(vi) Blood vessels: Increased permeability, odema of the vessels wall and surrounding tissues and haemorrhages.

(vii) Eye cataract.

(viii) Gonads – Sterility in both sexes.

(ix) Bone necrosis, osteosarcoma.

(x) Genetic effects: Still births failure to conceive, congenital malformation, mental defects, epilepsy, defects of neuromuscular, haematological and endocrine systems.

Therapeutic exposure of radiations to man is though harmless but sometimes creates readiation sickness which is characterised by nausea, loss of appetite, mild fever and malaise.

There is hardly any safe dose, slight increase in the background radiation level can cause cancer and leukaemia. It also induces mutation. The deleterious genes can persist in human beings, animals, plants and finally can affect their progeny.

The ionizing radiations such as alpha and beta radiations, neutron and gamma rays cause injury to the protoplasm and replace some elements (viz. calcium, carbon and zinc) when their concentration in the protoplasm increases.

Management of Radiation Pollution

The peaceful advantages of radioactive substances are so wide and effective that modern people cannot live without them. However, there is no permanent cure for preventing radiation damage, so all effects should be concentrated to minimize the radioactive pollution. The following methods can be used in preventing radioactive pollution:

(i) Leakages from atomic reactors should be regularly checked and must be controlled.

(ii) Careless handling, transport and use of radioactive fuels, fission products and radioactive isotopes must be totally stopped.

(iii) The safety measures should be fully and strictly enforced. For example if the amount of radioactivity is less than 5 millicuries, no precautions are essential. But if the body contains radioactive material in the range of 5 to 30 millicuries, then doctors must wear heavy rubber gloves, plastic aprons, plastic shoe covers and spectacle to reduce contamination.

(iv) Contaminated wearing dress should be thoroughly cleaned with soap and water, and stored for suitable decay of the radioactive material before sending to laundry.

(v) There must be a safe disposal of the radioactive waste material. Regular monitoring through frequent sampling and quantity analysis has to be ensured in risky areas.

(vi) Preventive measures have to be followed so that background radiation level does not rise above the permissible limits.

(vii) Safety measures should be taken and strengthened against accidents. Appropriate steps also require to be taken against occupational exposure.

(viii) Radioactive wastes having very low radiation should be discharged into sewerage.

Suggestions

(i) Radioactive wastes must be locked in insulated concrete filled sealed drums and, thereafter, be buried or discharged at least 1000 fathoms deep into the sea.

(ii) Wastes having low radioactivity should be stored for some time to further decrease their activity before their disposal.

(iii) The concentrated radioactive part of the wastes should be stored in special insulated tanks and their small quantities can be converted into solid such as concrete, glass or ceramic materials.

(iv) The decontaminated portion of the radioactive waste may be discharged but wastes having long half life period should not be discharged at all.

Impact Points to Remember for Radioactive Pollution

- Radiation is a part of man's environment.
- The sources of radiation to which man is exposed have been divided into two groups *viz.* natural and man made.
- Man is exposed to natural radiation from time immemorial.

- Natural background radiation arises from cosmic rays, atmospheric radiation and internal radiation.
- Man made sources of radiation are: X-ray, radioactive fallout and miscellaneous (T.V. sets, luminous wrist).
- Ionizing radiation refers to the radiation which has the power to penetrate into tissues and deposits its energy within them.
- Ionizing radiation are of two types *viz.* electromagnetic radiations and corpuscular radiations.
- Electromagnetic radiations consist of X-rays and gamma rays, and those of corpuscular compose of alpha particles, beta particles (electrons) and protons.
- Alpha particles are 10 times as harmful as X-rays, beta particles or gamma rays. But luckily they have less penetrating power.
- Alpha particles become quite dangerous if radioactive substance has entered the body by inhalation or through a wound.
- Gamma rays and X-rays have deep penetrating power owing to their short wavelength.
- Non-ionizing radiation refers to several forms of electromagnetic radiation of larger wavelength than those of ionizing radiation.
- Non-ionizing radiation includes ultraviolet (UV) radiation, visible light, infrared radiation.
- The potency of radiation is measured in three ways *viz.* Roentgen, Rad and Rem.
- Roentgen is the unit of exposure. It is the amount of radiation to absorb in air at a given point *i.e.* number of ions produced in 1 ml of air.
- Rad is the unit of absorbed dose. It is the amount of radioactive energy absorbed g^{-1} of tissue or any other material (1 ml rad – 0.001 rad).
- Rem indicates the degree of potential danger to health. It is the product of the absorbed dose and the modifying factors.
- The biological effects of ionizing radiation are two types *i.e.* somatic effects and genetic effects.
- Among the somatic effects radiation sickness, acute radiation syndrome, leukaemia, carcinogenesis, shortening of life and foetal developmental abnormalities are the most important.

- Genetic effects result from injury to chromosomes *i.e.* chromosomes mutations and point mutations.
- Managing of radiation pollution is the youngest branch of hygiene which is called radiation hygiene.
- The International Commission on Radiological Protection (ICRP), the International Atomic Energy Agency (IAEA), World Health Organisation (WHO) are very active in managing of radiation pollution.
- The ICRP's recommendations on radiation dose for occupational workers and for the general population have been adopted by many countries.
- The WHO (1993) has published permissible limits in drinking water.
- The IAEA has sponsored many symposia for disposal of radioactive wastes (Straub, 1970).

REFERENCES

Kotpal, R. L. and Bali, N.P. (1989). *Concept of Ecology*. Vishal Publications, Educational Publishers & Booksellers. Adda Hoshiarpur, Jallandhar City-6, 6 U.B. Bangalow Road, Delhi-7, India.

Park, K. (1997). *Park's Text Book of Preventive and Social Medicine*, M.S. Banarsidas Bhanot, Publishers 1167, Prem Nagar, Jabalpur, India.

Straub, C.P. (1970). Public Health Implications of Radioactive Waste Release, Geneva.

W.H.O. (1993). *Guidelines for Drinking Water Quality*, recommendations Vol. I, 2nd Edition.

11
FOOD POLLUTION

Introduction

Food occupies third position, after air and water for maintaining process of life on this planet. Resistance to disease depends upon the intake and purity of the vital components of food *viz.* proteins, carbohydrates, fats, vitamins, minerals and hormones. Absence or missing of any part of whole of the food constituents or addition of adulterants adversely affects of our health. The strength of a nation is, by and large, depedent upon the health and its population. The proverbs "Health is wealth" and "Sound mind in sound body" are well known to us. Therefore, purity of food and preventing contamination/adulteration by harmful ingredients are direct contributions to the National Health Development Programme. Food adulteration has long been enacted (1954) by the Government of India under which food adulteration is an offence and criminal act, and is liable to deterrent punishment. However, despite this act, the adulteration of the food has increased tremendously *vis-a-vis* pollution and contamination of food due to one or the other reasons.

Definitions of Food Pollution, Food Contamination and Adulteration

Food Pollution

Food pollution may be defined as the entry into food chain of undesirable substances through environmental agencies such as air, water and soil.

Adulteration

Mixing of undesirable foreign substances with food materials, for undue economic gain is known as adulteration.

Contamination of Food

The production of toxins in food through microbial agencies is known as food poisoning. The degradation in quality of food materials under the microbial influence to such an extent as to produce diseases on consumption, is termed as contamination of food.

Health Pollution

The sum total of all the pollutions is the health pollution. It is almost the same as the food pollution already mentioned above.

Causes and Hazards

The food pollutants, food contaminants and adulterants are of two types, chemical and biological. Both of these pollutants enter the food materials through environmental routes or by adulteration. The potential chemical pollutants found in food are fluoride, nitrates, pesticides, fertilizers, radioactive isotopes and a number of trace elements. Naturally present organic pollutants in some of the food are salanine, oxalic acid, erucic acid etc. Commonly used food adulterants are argemone oil, white oil, hydrogenated oil etc. toxins are produced by *Clostridium* bacteria, *Staphylococcus, Salmonella* and *Bacillus*. The detail for the above mentioned pollutants, adulterants, food additives and bacteriological toxins are briefly presented below:

Chemical Pollutants

Fluoride, nitrates, lead, cadmium, tin, mercury, selenium, aluminium, are some of the substances which cause pollution in food materials.

Fluoride

Tea, spice, gram, french beans and fishes are rich sources of fluoride, whereas vegetable and fruits are generally low in this respect. Among cereals, fluoride is maximum in wheat ranging from 0.8 to

1.0 ppm. People living in the regions where water contains more than 1.5 ppm of fluoride are advised to take citrus fruits and other foodstuff rich in vitamin C as they play an important role in preventing chronic fluoride intoxication. Excess of fluoride causes fluorosis disease. It results in mottled tooth enamel, abnormal calcification of bones and joints.

Nitrates

The nitrate pollution in drinking water is mainly through use of nitrate fertilizer, garbage dumps and animal wastes. High concentration of nitrate in water causes a serious disease, methaemoglobinemia (Blue Blood Baby) mostly in infants damaging respiratory and vascular system. It also leads to suffocation.

Lead

Presence of lead in air, water and food comes from exhausts of automobiles, gasoline contains tetraethyl lead. Its presence in human blood is more than 40 microgrammes per 100 ml blood can inhibit a number of enzymes which are essential for biosynthesis of haem (basic component of blood).

Recently tests carried by Bhaba Atomic Research Centre reveals that 10 per cent of Bombay city's population contain 30 microgramme of lead in 100 ml of blood. It was attributed to the presence of this metal in different foods.

Lead poisoning can do a havoc. It leads to blood losses, convulsions, liver and kidney problems and retardation in metabolic activities of the body.

Children in the habit of nibbling at iron or peels of old paints suffer from lead poisoning which adversely affects the body tissues. Though most of the lead is eliminated, some quantity does enter into the blood, bone and tissue.

There is malfunctioning of nervous system due to lead poisoning. In New York out of 2,018 children treated for lead poisoning, 128 died from 1954 to 1967.

Cadmium

Cadmium is not required by the human body and is, therefore, toxic even in low concentration. It has been reported to accumulate in human kidney and liver.

Cadmium occurs in air due to various industrial emissions and other activities of man. Refining, welding and electroplating processes, lead to pollution of water and air and from these sources it finally enters into human foods.

Cadmium accumulation leads to suppression of growth, bone deformation, renal problem, bloodlessness and anaemia, liver injuries, problems of malfunctioning of central nervous system gastric and intestinal disorders etc.

Excess of cadmium in food material (greater than 0.01 ppm) may create a negative calcium balance leading to osteomalacia (softening of the bones). Multiple fractures resulting from osteomalacia are responsible for the painful disease known as Itai-Itai. The cadmium intoxication can considerably be prevented by increasing the intake of vitamin C.

Tin

Oils and other fats, beverages, jams, fruits,vegetables, biscuits, coffee, tea etc, are marketed and stored in consumer houses in containers generally made of tin. Acidic foodstuffs are more prone to be contaminated by tin. Canned moist food on opening starts dissolving tin. For example, canned tins of apple juice when first opened may contain 20.3 mg of tin per kg of juice while after 2 days, the quantity may go up to 87 mg per kg and after 4 days upto 364 mg per kg.

The food stuff and beverages normally contain tin in amounts which are usually below 100 mg per kg and are generally well tolerated. The toxic dose in food which causes acute toxication is believed to be around 2 to 3 g per day for weeks together. The toxic effects of tin on human body have not been determined in much details as yet. Further studies may reveal still more horrifying details.

Selenium

Though the toxicity of selenium has been known for a long time its essentiality as a trace mineral for preventing retarded growth the reproductive troubles and specific disease known as "white muscle disease" came to be known very recently. The high incidence of this disease has been recorded in several parts of the world including India where soil selenium content is low. However, any soil containing above 0.5 ppm of the element is potentially dangerous and may cause selenium toxicity in grazing animals. The contamination of this element to human bodies is through water or milk and its products.

Selenium accumulation may cause, fever, nervousness, vomiting, dental caries, fall in blood pressure and even death. Liver, kidney and spleen may also be damaged.

Mercury

There are many well documented cases of discharges of mercury into various waterways. Mercury is many times heavier than water and it was assumed that waste mercury could be disposed of by dumping it safely into nearby lakes and streams. Previously it was not realized that mercury would undergo changes when introduced into water to form the compound methyl mercury. Methyl mercury enters into the food chain in the lower orders of the plants and animals and then advances into the marine food chain step by step and reaches our table. The adverse effects of mercury poisoning are well known.

Mercury deposits in human body lead to headaches, pain in some parts of body. Intestinal problems as diarrhoea, blood malfunctioning etc. About mercury, it is said that it leads to a disease called Minimata disease which is characterised by numbness and vision disorders.

Zinc

It is one of the essential element both for plants and animals. In animals, its deficiency is known to inhibit spermatogenic and testicular functions, and also cause para-karatosis disease. In human beings its deficiency causes dwarfism and pituitary hypofunction. However, excess of air borne zinc poses a threat to life. It occurs in the form of zinc oxide vapours having white colour. From bronze and brass refineries zinc may get accumulated to various kinds of food chain. Accumulation of zinc in human beings leads to kidney problems, pain in legs and even vomiting.

Iron

Iron is essential for formation of haemoglobin constituent of blood in human body, leghaemoglobin present in leguminous plants and ferratin compound. But excess of everything is bad, so is the case with excess absorption of iron. The excess iron is not excreted from the human bodies and causes disease called siderosis. This disease commonly occurs in those people, who always cook their food in iron utensils or iron pots. The disease siderosis leads to liver cirrhosis which is very fatal.

Other Metals

Other metals such as chromium, aluminium and copper have also been found to be harmful when present in foodstuff even in small amounts.

Chromium is said to be carcinogenic agent. It also leads to some kidney disorders, ulcerations in intestinal tract and some disorders in nervous system.

High levels of aluminium in human body weakens our nervous system and as such is harmful. Aluminium enters our body from aluminium cooking utensils mainly.

Copper has its own effects. It leads to mental stress and disturbance, coma and tension. It also results in uremia a disturbance when blood cannot dispose of some poisonous matter deposited in it.

According to tests conducted by Bombay Veterinary College, the arsenic concentration in some fishes sold in Bombay market was up to 2 ppm which is above the human tolerance. Arsenic in human bodies lead to liver damage, ulcers and some kidney problems.

Radioactive Elements

Man, among his various reckless activities, has been dumping different radioactive materials into the biogeochemical cycles. This radioactive mass comes to stay in the environment on account of chain reactions in nuclear weapon testing and in the nuclear reactors.

In this radioactive mass, there are many kinds of isotopes. Strontium 90, is one of them which has a half life period of more than 160 days. It is released as a product when uranium atom is split into smaller units alongwith cesium, barium and iodine. It is believed that strontium-90 is present in nuclear explosions upto the extent of 5 per cent and remains active for about 28 years. It behaves like calcium in biogeochemical cycles and is absorbed in living organisms. Grasses and vegetables absorb strontium-90 and through them it enters human body. There are reports that concentration of Strontium-90 arose in English milk by 250 per cent, one year after Bikni tests in March 1957 near Marshall islands in the Pacific Ocean. It is also found that Strontium-90 level is rising in children in some of these regions. Leukemia, bone cancer and some hereditary diseases are caused by strontium-90. The other affects of strontium-90 are birth abnormalities, deformities in children and other disorders. It has been reported that if nuclear tests are not stopped completely some 96,000 children world over are expected to be affected every year.

Use of Pesticides

The use of pesticides in agricultural production is poisoning potential health hazards not only to lilvestock and wildlife, but also to human beings, directly or indirectly through food chains. In India a survey was conducted during 1976 to find out the contamination of various foodstuffs with pesticides. For this purpose, various food commodities were collected from a number of cities *viz.* Hyderabad, Delhi, Mysore, Ludhiana, Pantnagar and from various sources *viz.* farmers houses, FCI godowns and markets. These were then analysed for pesticide. The results of this survey are summarised inTable11.1. The data of this table clearly indicates that most of the food commodities of India are contaminated by pesticides which have toxic effects on human beings.

Pesticide residues in food items, which are on the increase have become a matter of threat to man. Even small quantities of these residues ingested daily along with food, can build up high levels in the body fat. The long term effects of these residues in the human body include carcinogenecity, reduced life span and fertility, increase cholesterol, high infant mortality and varied metabolic and genetic disorders.

It is worthwhile to mention that 85-90 per cent of pesticides applied agriculturally never reach the target organisms, but disperse in the environment through air, soil and water. A number of studies carried out by different scientists in various parts of India, have detected residues of chlorinated hydrocarbon pesticides in soils, waters and all types of agricultural produce (Subba Rao, 1999).

The main source of dietary intake of DDT residues is through milk and milk products followed by oils and fats, in both vegetarian and non-vegetarian diets. The higher dietary intake of DDT residues through vegetarian diets, as compared to non-vegetarian diet, is due to the greater contribution of cheese in the diet. The dietary intake of HCH is also mainly through milk and milk products followed by meat and eggs for the non-vegetarian diet, whereas it is through cereals followed by milk and milk products of vegetarian diet (Tables 11.2 and 11.3).

Table 11.1. Pesticide Contamination of Food Samples collected from Delhi, Hyderabad, Mysore, Ludhiana and Pantnagar

Commodity	No. of sample examined	No. of samples found contaminated	Pesticides detected
Wheat grains	659	190	DDT, BHC, Malathon
Wheat flour	2	2	DDT
Rice	4	4	DDT
Cereals	77	56	DDT, BHC
Pulses	32	16	DDT, BHC
Vegetables	727	514	DDT, BHC, Endrin, Chlordane and Heptachlor
Milk	15	11	DDT, Endrin
Butter	2	2	DDT
Eggs	21	14	DDT, BHC
Fat of baffalo, goat and cows	63	61	DDT

Source: Gupta, R.D. (2000). Compiled from different books and journals.

Table 11.2 : Pesticide residues detected in various items

Item	No. of samples analysed	Percentage of samples containing residues
Soils	706	84
	359	93
Waters	109	3
	40	100
Wheat	1085	38
Rice	-	81
Sorghum	-	38
Pulses	487	76
Vegetables	4,801	52
Fruits	-	80
Spices	42	100
Honey	-	62
Milk	1,191	67
Butter	-	66
Ghee	5	100
Infant	186	70
Vegetable oils	92	53
Fish	14	50
Egg	12	83
Human Milk	130	100
Human tissues	682	92

Source: Subba Rao, (1999). *J. Indian Soc. Soil Sci.* 47 : 611-633.

Table 11.3: Estimates of daily intake of DDT and HCH through consumption of milk contaminated at mean level in different States in India

State	Intake (μg day^{-1})			
	Adult		Child	
	DDT	HCH	DDT	HCH
Andhra Pradesh	32.5	142.5	65.0	171.0
Bihar	10.0	44.5	12.0	53.4
Gujarat	22.5	17.7	27.0	21.3
Haryana	5.0	12.2	6.0	14.7
Himachal Pradesh	22.5	8.2	27.0	9.9
Karnataka	12.5	48.7	15.0	58.7
Kerala	7.5	19.7	9.0	23.7
Madhya Pradesh	10.0	12.7	12.0	15.3
Maharashtra	20.0	11.7	24.0	14.7
Punjab	27.5	16.5	33.0	18.8
Uttar Pradesh	7.5	59.7	9.0	71.7
West Bengal	5.0	42.5	6.0	51.0

Source: Singh & Dhaliwal. (1993). In *Pesticides* (Dhaliwal & Singh, Ed.), Commonwealth Publishers, New Delhi.

The data regarding pesticide residues identified in various food, fodder and feed items collected by Acharya N.G. Ranga Agricultural University scientists over the years (Table 11.4), indicated presence of high and toxic amounts of pesticide residues in foodgrains, fruits, vegetables, milk and milk products, eggs, soils and waters. Even human breast milk is reported to contain pesticide residues. The toxic effects of some of the pesticides on mankind is as follows:-

DDT and BHC

These pesticides do not decompose readily and leave intolerably high residues in the soil, water and remain in fruits, vegetables, grains etc. On being consumed these contaminated foodstuffs by livestock and human beings deliver deformed babies or have still births. Human DNA is also adversely affected. DDT causes cancer besides a number of other defects in human beings. Kant (1987) has reported that there are about 37,75,000 cases of human poisoning by pesticides in developing countries every year, with some 10,000 deaths. He further reported that the level of accumulated DDT in the body tissues of the average Indian is the highest in the world. About 70 per cent sample of bottled milk in Maharashtra contained 4.8 to 6.3 ppm DDT and 1.9 to 6.3 ppm dieldrin. The permissible limit of these two compounds is 0.66 ppm. Even baby food samples collected from Punjab, Gujarat and Bombay had DDT residues ranging from 1.52 to 2.12 ppm.

The lethal effect of DDT, BHC and aldrin *vis-a-vis* other insecticides on various forms of wildlife has been reported by Malhotra and Sharma (1987) in greater detail.

Parathion/Malathion

These are organophosphates and affect the nervous cells of human being. Muscular degeneration and dizziness are the other effects of these substances.

Carbamates

Carbofuran, zectran and isolan are the main insecticides included in this group. These also get accumulated in food chains at various levels adversely affect nervous system, muscular co-ordination and other disorders.

Rodenticides and Molluscicides

Rodenticides like warfarin, zinc phosphide and fluoroacetate are used to control rodents, whereas molluscicides are of common use to kill the molluscs, snails and slugs. No doubt, these chemicals reduce the population of these pests but application of these substances seriously interfere with structure of aquatic vegetation.

Table 11.4 : Pesticide residues identified in various food, fodder and food items

Pesticides	Items of food having residues of tonic levels
HCH	Rice, milk, egg, tomato, carrot, cucumber, onion, radish, brinjal, bhindi, bitter gourd, green chillies, dry chillies, chillies powder, paddy grains, rice bran, paddy straw, water samples, rice growing soils. Certain brands of cooking oils, milk, milk products like skim milk, milk powder, butter, cheese, infant milk powders. Poultry feed – maize, cake, bran, fish, feed mix. Livestock feed – eggs, fish, corn cakes, corn+cakes, wheat + jowar+groundnut cake.
DDT	Cooking oil, milk, butter, maize, fish, feed mixture, livestock feed, eggs, human breast milk.
Monocrotophos	Tomato, onion, cooking oil one brand
Cypermethrin	Cooking oil – certain brands
Quinolphos	Fish, cooking oil – one brand
Aldrin	Maize, fish, feed mix, livestock feed, eggs, human breast milk
Endosulfan	Livestock concentrated feed, fish

Source: Annonymous. 1994. Pesticide Contamination – Bull.

The jaundice epidemics of Baramula and Sopore (Kashmir) is traced back to spraying of apple trees with deadly chemicals against scab which find their way in drinking water supplies. There was loss of so many human lives.

The diarrhoea epidemic in Rajouri-Poonch areas (Jammu, J&K) during 1986 took toll of many human lives and it was reported that water was polluted with diarrhoea causing germs.

Estimates of daily intake of pesticides and their residues in various foodstuffs determined recently (Singh & Dhaliwal, 1993 and Anonymous 1994) have been shown in Tables 11.3 and 11.4.

Use of Fertilizers

With the introduction of high yielding varieties of various crops, the use of fertilizers has been tremendously increased. The fertilizers comprise one or more of major plant nutrients, mainly nitrogen, phosphorus and potassium. The fertilizers applied to the soil may be carried to ground water by leaching. The nitrates are more soluble in water and as such leach to ground water readily. The fertilizers can also mix with surface water by run off and drainage. In lakes and reservoirs, the inorganic nitrates and phosphates stimulate excessive plant growth which is harmful to fish and other aquatic life. Excessive amount of nitrates (> 45 ppm) in ground water is also harmful to human beings. Its harmful effect on human beings has been mentioned in water and soil pollution chapters.

Fertilizer industries discharge effluents with high content of ammonia (100 ppm), urea (180 ppm), phosphate (40 ppm) and sulphate (1200 ppm), which subsequently find their ways into the adjoining water courses, causing bloom of algae. Ammonia in concentration to 0.1 ppm in rivers, streams or lakes is toxic to fishes. The fertilizer waste is usually accompanied by arsenic and chromium and unfortunately both of them are equally objectionable for aquatic fauna and flora. The wastes generated from alkali manufacturing industries contain mercury, which builds up 1000 times higher concentration in fishes than found in receiving water course. The Minimata disease of Japan is well known example of mercury hazard where thousands were poisoned and hundreds died by consuming mercurised fishes. The phosphate pyrite mining produces wastes of very high acidic nature, accompanied by as low as pH value of 1.

Application of fertilizers, insecticides, weedicides and growth hormones has assumed alarming proportion. Our soils are thoroughly polluted and they produce polluted agricultural products. The use of chemical fertilizers has weakened the natural vitality and vigour of our agricultural lands. The microbiology of the soil has been destroyed. The earthworm population have declined and the physical and biological balance of the soil has been upset.

Adulterants

Argemone oil

The argemone seeds are black in colour like those of mustards and the latter is, therefore, very much adulterated with the former. It has been found that regular consumption for 1-3 months of any edible oil, contaminated with as low a concentration as 0.1 per cent of argemone oil can give rise to epidemic dropsy in human beings. Dropsy is a disease causing swelling of feet, thighs and face. If the consumption of adulterated oils is not discontinued, it may cause glaucoma (disease of the eye) leading to blindness or swelling of liver which may be fatal.

Tricrecyl phosphate (TCP)

TCP is practically colourless, odourless, non-volatile and stable liquid, insoluble in water but is freely miscible in vegetable oils. It is used as adulterant with the mustard oil. Consumption of oils containing TCP for at most 20 days may cause flabby paralysis.

Vegetable oil (Vanaspati)

Vanaspati unlike ghee (butter oil) contains trans acid which may cause adverse effects on the consumer's health on prolonged use especially when it is adulterated with ghee.

Microbial toxins

Clostridium botulinum

It causes food poisoning and occurs mostly in improperly sterilized preserved foods which have a near neutral or slightly alkaline condition. The poison caused by this bacterium is potent because ingestion of even minute quantity would result in a severe syndrome that may be fatal within 2 to 10 days.

Staphylo-coccus poisoning

Staphylo-coccus grows in cream, puddings, sauce and other starchy preparations. Besides these, milk, egg and meat are also prone to their attack especially during summer. However, in contrast to salmonellosis or botulinum the attack does not last more than one day, and usually wanes within 6 hours or less.

Natural Organic Toxicants

Solanine

It is a group of toxic glyco-alkaloids found in a number of plants. Potatoes damaged or black potatoes may contain upto 400-500 ppm of this toxicant. Consumption of such potatoes may cause abdominal pain and diarrhoea.

Gosypoll

It is a toxicant made up of phenolic compounds and may be found in free state as well as in combination with protein in cotton seed and its flour, but never in cotton seed oil. It is toxic to man, pig, sheep, chicken and fish but not to the ruminants (cows and buffaloes). The maximum allowable concentration of this toxicant is 1.1 per cent in cotton seed flour used for making biscuits.

Oxalic acid

It is present in Chana dal (180 mg), amla (296 mg) and lathyrus oil (122 mg) per 100 g. Consumption of heavy dose of this food produces acid cramps, haemophilia and even death. It is advisable to restrict the intake of foods containing more than 0.5 per cent of oxalic acid (500 mg/100 g of food material), incidence of information of stones in kidney is high.

Erucic acid

It is a long chain unsaturated fatty acid normally present in mustard oil mixed with rape seed in concentration exceeding to about 45 per cent. Habitual use of such mixed oil has been found to cause myocardial fibrosis (reduction in eye sight in human being).

Amygdalin

It is a glycoside. It is composed of glucose, benzaldehyde and hydrocynic acid and mainly occurs in bitter almonds. It has been found that continuous consumption of 12-16 bitter almonds by human beings, cuase giddiness and prove fatal.

Control Measures

A primary objective for checking food pollution is to control soil pollution. The methods to check soil pollution have been discussed in "Soil Pollution" chapter. Further to this discussion, a few relevant points are added here:

(i) Discover plant species which have capacity to absorb different pollutants for instance, pines are known to absorb atomic fallout efficiently (Jamwal, 1987). *Bauhinia* and Eucalyptus can absorb high quantities of sulphur dioxide. Willows and Ailanthus are good absorbers of methane (CH_4) and hydrogen sulphide (H_2S) of the city wastes.

(ii) Grow more trees and shrubs to have more foliage and more oxygen.

(iii) Create awareness among the urban and rural masses about the importance of hygiene and sanitation of our cities, towns and villages. More filth and dust means more food pollution and toxicity.

(iv) Prior to cooking the vegetables, these must be washed twice or thrice with warm water. This will reduce the contamination of the pesticides. Similarly, all fruits before eating must be thoroughly washed.

(v) Safety glasses and protective clothings are some of the control measures which can be employed to reduce food contamination. For example, on cafeteria lines glass sneeze guards must be used to profect against food contamination. Lead may be used to restrict the emission of stray ionizing radiations from X-ray machines.

(vi) To reduce the bacterial contamination cooking and boiling of foodstuffs is very essential. Wherever possible, autoclaving and pasteurization should be followed to kill pathogenic or other undesirable microorganisms.

(vii) Minimizing exposure through restriction or activities, immunization against the disease and the use of prophylactic agents may avoid infections and, as such, are measures to be used to avoid the effects of environmental hazards. One may use only boiled water and cooked food when travelling in areas where drinking water is not safe or sewage is used to irrigate the growing vegetables. The avoidance of unpasteurized milk and uncooked shell fish are the other practices which should be followed by the travellers.

Priorities and Suggestions/Future Research Work

(i) Research on organic farming/organic gardening is the need of the day in the light of creating soil pollution by using chemical fertilizers.

(ii) Research on natural control of pests and disease should be taken up on large scale. For this purpose, evolve such plants which can kill the pests and can also kill the unwanted plants (weeds). This strategy will enable us not to use pesticides and weedicides.

(iii) Develop more resistant strains and varieties of our agricultural crops against insects and pests.

(iv) Try to produce natural immunity in various crop plants against diseases through selective cross breeding. Evolve superior genetic materials.

(v) Discover plants which have specific natural healing capacity. Scientists of Russia have discovered that fir forest heals high blood pressure and heart patients effectively.

(vi) Use of colours in food preparation like confectionery, halwas and other sweetmeats, soft drinks, condiments and spices, should be totally banned to prevent food poisonings.

(vii) Food adulteration acts must be enforced strictly. Any person who violates any provision of acts, should be penalized or punished with heavy fine or imprisonment.

(viii) The work of food preservation technology, processing of food and packings, and treatment of dry fruits, should be handed over to scientists and specialists rather than industrialists. The scientists/specialists should be given free hand to tackle the food hazards/problems.

(ix) Find out root causes of food toxicity and try to remove the cause naturally rather than to use chemical antidotes.

Impact Points to Remember for Food Pollution

- Today's industrial world has introduced thousands of pollutants and poisons into our food.
- Scientists have identified five principal causes of food pollution: (i) bacterial contamination, (ii) naturally occurring toxins, (iii) chemical carcinogens, (iv) pesticide contaminant, and (v) food additives.
- Poor food processing techniques and bad packing also contribute to the rising food poisoning cases.
- A survey conducted has revealed that bottled milk supplied in most metropolitan centres in India, contained almost ten times more DDT than the permissible limit.
- Numerous food additives used for flavouring colouring, preserving and as emulsifiers and sweeteners, add nothing to the nutritive value of food but instead cause health problems. For example, oil and fat preservatives can remain in the body for several years releasing toxic substances and affecting metabolism.
- In food processing, plastic containers used are made from a variety of noxious chemicals which leach into the food.
- Chemical additives have been found to produce a variety of ailments, ranging from gastroenteritis to cancer as well as damages to kidney, liver and nervous system.
- The chocolate in icecream may contain very little cocoa but enough of amyl-phenylacetate, vanillin and propylene glycol to cause digestive disorders.
- Many sweets contain carboxymethyl cellulose which is now considered to be carcinogenic.
- There is always the risk of pathogenic micro-organisms entering the body through water.
- Heavy contamination of river and waterways have always resulted in outbreak of epidemics like enteric fever, dysenetry, cholera, diphtheria and poliomyelitis.
- Recently the Indian Council of Medical Research conducted a survey in 12 States and found that foodstuffs were adulterated with harmones. For example the oxytocin injection given to animals for higher milk yield, is secreted into milk also.

- The dropsy epidemic is still fresh in memory. That was due to mustard oil adulteration.
- The Government of India has set up a committee to fix standards on various food items and to prepare guidelines, for controlling food adulteration.
- There are only 72 laboratories for samples examination in the country, although the aim was to have one in every divisional headquarters.
- As far as the appellate laboratories are concerned, only four exist to satisfy the adulteration act, situated in Ghaziabad, Kolkata, Pune and Mysore.
- Grow more trees and shrubs for absorbing different pollutants to prevent their entry to food chain.
- Prior to cooking, the vegetables must be washed twice or thrice with warm water to reduce the contamination of pesticides and other pollutants.
- Safety glass and protective clothings are some of the control measures which can be used to lessen the pollution.
- To reduce the bacterial contamination boiling of foodstuffs is necessary.
- Certain laws must be framed by the Government to check food adulteration.

REFERENCES

Anonymous (1994). *Pesticide Contamination Bulletin.* ANGRAU, Hyderabad.

Jamwal, G.S. 1987. Toxicology of Pollution. In Environment & Pesticide Toxicity. Proceed. 8th Annual Session of the Academy of Environmental Biology, University of Jammu, J&K, India, pp 23-31.

Kant, S. 1987. Environmental Pollution by Pesticides. In Environment & Pesticide Toxicity. Proceed. 8th Annual Session of the Academy of Environmental Biology, University of Jammu, J&K, India, pp. 97-111.

Malhotra, Y.R. and Sharma, K.K. (1987). Pesticides and Wildlife. In environment & Pesticide Toxicity. Proceed. 8th Annual Session of the Academy of Environmental Biology, University of Jammu, J&K, India, pp. 147-160.

Singh, B. and Dhaliwal, G.S. (1993). *In Pesticides : Their Ecological Impact in Developing Countries.* (Ed. Dhaliwal G.S. and Singh, B.). Commonwealth Publisher, New Delhi, pp. 131.

Subba Rao, I.V. (1999). Soil and Environmental Pollution – A threat to sustainable agriculture. *J. Indian Soc. Soil Science.* 47(4) : 611-633.

GLOSSARY

Abatement : A control measure taken for eliminating or reducing air, water or noise pollution is called abatement. It may involve legislative and technological applications.

Abiotic : It indicates non-biological component and is a physical or chemical feature of an environment or ecosystem.

Absorption : It is the phenomenon by which a substance accumulates on the surface of another substance and goes into the interior of the surface also.

Acacia : It is a genus of leguminous trees of tropical or subtropical origin.

Acid : A substance containing active hydrogen, that dissociates on solubilizing in water to produce hydrogen ions, is called acid or acid is a substance that tends to donate proton during chemical reaction.

Acid forming fertilizer : Fertilizers which on application leave an acid residue in soil.

Acid rain : Precipitation, whether rain or snow wherein water possesses acidity greater than normal, which is due to an occurrence of sulphuric acid and nitric acid.

Acidification : Soil forming process by which acid soils are formed.

Acidity : It denotes hydrogen ion concentration or activity of any substance, which is usually called pH. pH of pure water is 7 and below this figure, the water becomes acidic.

Additive : Any material added to improve its flavour, colour or physical condition of any food stuff or fertilizer is called additive.

Adsorption : Surface phenomenon by which a substance accumulates on the surface of another substance.

Adulterant : Any material, which is used for adulteration of the food material or fertilizer, is known as adulterant.

Adulteration : It represents mixing of any deleterious or harmful ingredient present in sufficient amount in food or fertilizer, which becomes injurious to human, animal and plant life.

Aeration : Addition of air to sewage or water for raising its dissolved oxygen level, is called aeration.

Aerobe : An organism which can grow and live only in the presence of oxygen.

Aerobic : Living or active object which thrives in the presence of free oxygen. It is the normal condition of a healthy *Azotobacter or Bacillus.*

Aerosol : It refers to a dispersion or suspension of a very small particles or liquid droplets or both present in the air.

Agricultural wastes : Any refuse of any form and kind which is generated from different agricultural operations, is an agricultural waste.

Air pollutant : Any foreign substance or material which pollutes the air, is called the air pollutant.

Air pollution : Contamination of the air with undesirable solids, liquids and gases or addition of any foreign substance that disturbs the normal composition of the air, constitutes air pollution.

Algae : A group of unicellular or multicellular simple plants containing chlorophyll, belong to soil organisms or other bodies.

Alkali : Hydroxides and carbonates of alkali metals viz., lithium, sodium, potassium, rubidium etc., having pH more than 7.0 of their aqueous solutions are known as alkalies.

Alkalization : It is a soil forming process by which alkali or sodic soils are formed.

Allergy : A specific substance that sensitizes human beings and animals, when exposure of the sensitive host provokes over illness.

Ammonification : A biochemical process which produces ammonical nitrogen from organic compounds having nitrogen.

Ammonifying bacteria : Those bacteria which break down the proteins or allied substances into ammonia are called ammonifying bacteria.

Amoebiasis : (Amoebic dysentery) : It is a disease caused by a protozoa, *Entamoeba histolytica.*

Anaerobe : An organism which thrives in absence of oxygen, is called anaerobe.

Arsenic : Element 33. Arsenic pollution is often associated with the manufacture or use of herbicide / pesticide.

Asbestos : A broad term which embraces several fibrous minerals.

Asbestosis : It is a disease of lungs which is caused by inhalation of asbestos particles.

Ascorbic acid : Vitamin C.

Atmosphere : The gaseous envelope over the earth's surface is the atmosphere.

Azotobacters : These are free living bacteria, which are capable of utilizing atmospheric nitrogen under aerobic condition.

Bacillus : Any rod-shaped bacterium that produces endospors and thrives under aerobic conditions, is known as *Bacillus.*

Bacteria : These are the smallest and most primitive unicellular microorganisms, lacking of chlorophyll and multiply mostly by fission.

Bacterial Culture : Any media, when is enriched with any particular bacteria, is called bacterial culture e. g. *Rhizobium* culture, *Azotobacter* culture.

Bactericide : Anything which kills bacteria.

Bacterium : Singular of bacteria.

BGA : Blue green algae which belong to the class Cyanophyceae or Myxophyceae.

Biocide : Any agent that kills living organisms biologically.

Biofertilizer : Any biological material which is used as a source of supplying plant nutrients through various biological reactions in soils/plants, is a biofertilizer. *Azotobacter, Rhizobium, Phosphobacterin,* Blue green algae, *Azolla* are the main biofertilizers.

Biological control : It is the control of pests by the use of living organisms.

Biological half life : It is the time required for the body to eliminate one half of an administered dose of any substance by regular process of elimination.

Biological nitrogen fixation : Atmospheric nitrogen fixed by the living organisms for making its availability to plants is called biological nitrogen fixation.

Biotic : It pertains to life.

Botulism : A disease that is produced by *Clostridium botilinum.*

Bromofluorocarbons : These are agents which are responsible for causing depletion of ozone layer.

Cadmium : It is an element. Cadmium pollution is blamed for the Japanese "itai-itai" disease.

Carbon monoxide : An invisible, tasteless, odourless and highly poisonous gas, is found in many industrial areas and also as an emission from vehicles.

Carboxyhaemoglobin : It is a combination of carbon monoxide and respiratory pigment occurring in the blood plasma haemoglobin.

Carcinogenic : Cancer producing.

Carnivore : An animal or plant feeding on and digesting animal substance as a source of food.

Catchment : A drainage area or basin which collects all the rainwater falls on it and drains into a river or stream.

Catchment area : Any area from which water drains to a particular location such as main river system or a lake.

Chlorinated hydrocarbons : Organic compounds consisting of chlorine, hydrogen and carbon e. g., DDT.

Chlorofluorocarbons : These are compounds of chlorine, fluorine and carbon which like bromofluorocarbons, are responsible for causing depletion of ozone layer in the atmosphere.

Cholera : A serious intestinal disease which is caused mainly by *Vibrio cholerae.*

Chromium : White metallic element which is very toxic in its hexavalent form.

Compost : Mass of rotten organic matter which is mostly obtained from plant residues (leaves, stalks, twigs, bark etc.) with some animal dung.

Composting : A biological process by which compost is prepared under microbial decomposition.

Conservation : Preservation or rational use of any natural resource for the posterity.

Contamination : A general term which signifies the introduction of microorganisms, chemicals or wastes into air, water, soil and food, rendering them unfit for use.

DDT (Dichloro-diphenyl-trichloroethane): Persistent organochlorine insecticide.

Decibel : Unit of noise. Nearly all audible sounds lie between 0 and about 140 dBA.

Deforestation : Felling of trees of forests for arable land or to use the timber for construction or industrial purposes.

Denitrification : The breakdown of nitrates by soil bacteria resulting into the release of free nitrogen.

Disinfection : The killing of the harmful and objectionable microorganisms in or on a medium by means of chemicals, heat ultraviolet light etc.

Diversification : An increase in the variation exhibited by a population or community over time.

Ecosystem : The word ecosystem is derived from two words, "ecology" and "system"; the ecopart of the word implies environment while system of the word implies an interacting interdependent complex. So ecosystem is an interaction between physical and biological environment.

EDTA : Ethylene diamine tetra acetic acid. It is a chelating agent.

Effluent : A general term denoting any fluid which is emitted by sewage water or other liquids partially or completely treated or in its natural state flowing out of a reservoir, basin treatment plant.

Emission : The total amount of solid, liquid or gaseous pollutant emitted into the atmosphere from a given source (say, for example, automobiles) in a given time and indicates in gram per cubic metre of gas.

Endangered species : Plants and animals, which are likely to become extinct.

Endemic : (a) it is confined to a given region having been originated there; (b) Of pests or disease producing species. A species which continuously occurs in a given area.

Endemic disease : Disease that constantly recurs in a particular locality.

Endotoxin : A toxin released by the degeneration or death of bacterial cells.

Environment : The region, surroundings or circumstances in which anything exists.

Epidemic : A disease that spreads quickly.

Erosion : A process by which rock/soil particles are detached from their original sites, transported and then deposited to a new place or locality.

Escherichia coli : A bacterium of exclusively faecal origin which forms 90% of coliform group of bacteria in the human intestine.

Eutrophication : The ageing of a body of water by the growth of vegetation particularly algae which is brought about by an accumulation of nutrients (usually N&P).

Evergreen : A plant or tree which bears leaves throughout the year *e.g.*, pine tree.

Fauna : Animals of a particular region.

Fertilizer : It is defined as the material added to the soil or applied directly to crop foliage to supply elements for plant's growth.

Flocculation : The process of converting a finely divided or colloidally dispersed suspension of a solid into particles of such size in which rapid settling occurs.

Flora : These are plant species of a region or it is the plant life of a geological period.

Fluorosis : It is the disease caused by over consumption of fluorine compounds.

Flyash : Very small, powdery and glassy like material is called flyash, which is carried away usually from a coal fire.

Fog : It is a collection of tiny water droplets that float in the air. It is similar to clouds except that clouds do not touch the earth's surface as fog does.

Food additive : The food additive means any substance which is added intentionally to food generally in small quantity to improve its flavour, texture and appearance.

Fossil : Remains or traces of an organism that have been preserved in earth's crust by natural processes.

Fungi : Any tiny aerobic, heterotrophic Protists containing no chlorophyll, many celled, is called fungus (plural : fungi).

Fungicide : Any pesticide which inhibits or prevents fungus, causing diseases in plants, is called fungicide.

Garbage : Animal, plant, fruit or vegetable residues resulting from the handling, preparing and cooking of food, is known by the term 'garbage'.

Gastroenteritis : Inflammation of the stomach and intestine or enteric fever is designated by gastroenteritis.

Glaciers : A body of ice originating on land by compaction and recrystallisation of snow and showing evidence of present or past movement.

Grassland : Land which is dominated by herbaceous vegetation and grasses of various species.

Green manure : Any green crop that is buried directly into the soil to increase its fertility and to improve physical condition and chemical properties is called green manure.

Habitat : Any physical part of environment which is inhabited by an organism or population of organisms, is known as habitat.

Haemoglobin : A respiratory pigment that is found in the blood plasma.

Hail stone : Precipitation in the form of small balls or irregular-shaped lumps of ice.

Half life : The time required for the body to eliminate one-half of an administered dose of any radioactive substance by regular process of elimination.

Herbicide : Any chemical substance that is used to destroy off-plants or weeds.

Humification : It is process by which organic matter is transformed into humus.

Humus : A complex organic component of the soil, resulting into from the decomposition of plant and animal tissues, which is colloidal; amorphous; dark brown or black brown in colour and resistant to further decomposition.

Inceneration : A method of treating refuse to reduce the volume and the weight and to have an innocuous residue.

Indigenous : Native or original to an area, not introduced from outside, the particular region or environment.

Inhibitor : The opposite of catalysts as it retards or reduces rate of a chemical reaction.

Inoculation : Treatment of soil or seeds with bacterial culture, is known as inoculation.

Inoculum : Bacteria when are added in compost or soil to start biological action.

Insecticide : Any pesticide that kills insect is called insecticide.

Kiln : A furnace in which heating operations do not involve fusion.

Lake : It is a large non-flowing body of water held in depression in the earth's surface without direct access for mixing with the ocean.

Landfill : The disposal of refuse by tipping it on land.

Lead : Element, it is poison whose compounds accumulate in the body.

Lichen : It is an association of blue green algae or green algae with fungi.

Lodging : Falling of crop plants under the influence of heavy rains or wind.

Mercury : A silvery liquid, metallic element which is highly toxic.

Methane : The simplest hydrocarbon (CH_4), natural product or liberates through anaerobic decomposition. In natural form it is also known as marsh gas.

Microbe : Any tiny plant or animal which can be seen only with the help of microscope, is called microbe.

Microfauna : Microscopic animal *e.g.*, protozoa.

Microflora : Microscopic plant life *e.g.*, bacteria.

Micronutrient : An element which is required by the plant in small quantity.

Microorganisms : Microoraganisms are either microscopic plants (bacteria, actinomycetes, fungi) or animals (protozoa, nematodes).

Mining : It is the process of removal of metal ores, coal or other minerals from surface or underground deposits.

Natural pollutant : A substance of natural origin that may be considered as environment pollutant when exists in excess amount *e.g.*, ozone formed photochemically, dust and gases of volcanic origin.

Night soil : Human excrement or excreta.

Nitrate : It is chief source of nitrogen by which the plants take their nitrogen *i.e.*, in the NO_3 form.

Nitrogen oxides : It is one of the most important group of atmospheric contamination, like NO, NO_2 etc.

Noise : It is unwanted sound.

Noise pollution : When noise becomes intolerable and harmful to health, is called noise pollution.

Organic : It relates to a huge class of compounds containing carbon including those present in or derived from plants or animals.

Organic farming : Farming, which is done without the use of artificial or chemical fertilizers and pesticides.

Organic matter : The term "organic" refers to "living" so organic matter is any material derived from living organism like plant or animal.

Organism : Any living thing is designated by the term 'organism'.

Ozone : A highly reactive form of oxygen (O_2), is produced naturally during lightening storms by the passage of electricity through air. It occurs in stratosphere as ozone O_3 layer and absorbs the ultraviolet rays.

Parasite : Any organism that obtains its food from another organism. Normally on which it lives for part or whole of its life, is called host. Sometimes the parasite kills the host.

Particle : A small discrete mass of solid or liquid such as in aerosols, smokes, dusts etc.

Particulate : A particle, whether liquid or solid, especially in smoke which includes aerosols, fume, grit etc. The level of particulates in air is reported in microgram or total suspended particulates per cubic metre.

Pathogen : A microbe that causes disease.

Pest : Any plant or animal that, in its location, is an economic, aesthetic, physical or biological threat to humans or their possessions. Insects and weeds are two largest groups of pests.

Pesticide : Any chemical which is used to kill the pest, is called pesticide or which is used to control the plants and animals from an assault of pests.

Petroleum : A naturally occurring complex mixture of liquid hydrocarbon compounds is called petroleum which contains 82-87 per cent carbon and 11-15 per cent hydrogen.

pH : It is an extent of alkalinity and acidity in aqueous solution. Technically it indicates a negative logarithm of hydrogen ion activity.

Pollutant : Any introduced substance that adversely affects usefulness of a resource.

Pollution : A term which is used to indicate pollutant or a substance into the environment is potentially harmful or that interferes with man's use of environment.

Radiation : It is an emission and propagation of energy through space or material medium in the form of waves.

Radiation ionising : Any electromagnetic or particulate radiation which is capable of producing ions directly or indirectly, in its passage through matter.

Radiation pollution : Pollution, which occurs mostly through fallout from atmospheric test of nuclear weapons or atomic energy which accidentally release radioactive elements.

Recharge : Process or renewing underground water by infiltration during rainy or wet season is called recharge.

Recycling : In plant nutrition, it refers to putting back to soil the nutrient removed by the crop, using organic urban and rural wastes. In popular usage it is reclamation of waste product to manufacture a new product.

Rhizobium : These are bacteria which have the ability to fix atmospheric nitrogen in association with legumes.

Salinity : The amount of salts dissolved in water is called salinity or amount of salts present in the soil greater than 0.20 per cent is called soil salinity. Salt water contains less dissolved oxygen than fresh natural water.

Salinization : It is soil forming process by which saline soils are formed.

Sanitary landfill : A method of waste disposal by spreading the wastes over land and covering them with a seal of earth or an engineered burial of refuse into trenches and compacted by a bulldozer for its decomposition by microorganism to stable compounds.

Sludge : Solids settled out from sewage containing about 55-99 per cent water. Sewage is the waste disposal of industry or towns carried off in the sewage system.

Smog : It is a combination of smoke and fog.

Smoke : A dispersion of fine divided particles as gaseous material in air produced mostly by combustion.

Sodicity : When soil contains more content of exchangeable sodium. Usually sodic soil possesses exchangeable sodium more than 15 per cent of the cation exchange capacity.

Soil profile : A vertical section of soil is called soil profile.

Solanine : A naturally occurring poison found in plants belonging to the family Solanaceae.

Solid waste : Anything solid in nature which is discarded. Where part or all of the substance will be recycled-rubbish, trash. sewage sludge etc.

Spore : A seed or reproductive but dormant cell.

Stratosphere : The upper layer of the atmosphere lying above troposphere.

Toxicant : A chemical that controls pests by killing rather than repelling them.

Toxin : A poison which is produced by an organism.

Typhoid fever : It is the most serious enteric infectious disease. Infection is through food or water.

UNEP : United Nations Environment Programme.

Vitamin : An organic substance which is needed in very small quantity for human beings and other organisms especially belonging to heterotrophic group, to maintain growth and development.

Waste : Any matter, whether liquid, solid, gaseous, radioactive which is discharged in the environment, constitutes waste.

Water pollution : Substances like bacteria or viruses or other pollutants which impair the quality of water, rendering it less suitable for use, is called water pollution.

Wildlife : A collective term embracing several thousand different species of animals, birds and reptiles, is called the wildlife.

BIBLIOGRAPHY

Anonymous, 1992. Neem – a tree for solving global problems. National Academy Press, Washington D.C.

Anonymous. 1999. Water : Vision 2050. *Bhagirath.* 46 (4) : 36-45.

Das, S.K. and Bhadur, J. 1999. *The Himalayan Environment* . New Age International Limited Publishers. New Delhi. Bangalore, Kolkata, Mumbai, Chennai. Guwahati, Hyderabad, Lucknow.

Gopalan, C.; Kamala, J.R. 1980. In : *Prevention in Childhood of Health Problem in Adult Life* (Ed. Falkner. F.) Geneva, WHO.

Gopalan; C.; Ramasastri, B. V. and Balasubramanian, S.C. 1985. National Institute of Nutrition, Indian Council of Medical Research, Hyderabad, India.

Government of India. 1999. National Child Survival and Safe Motherhood Programme, MCH Division, Ministry of Health and Family Welfare, New Delhi.

Khosla, P.K. and Sehgal, R.N. 1988. Trends in Tree Sciences. Indian Society of Tree Scientists, Department of Forestry, Y.S. Parmar University of Horticulture and Forestry, Nauni Solan-173230, H.P., India.

Motsara, M.R. and Singh, J. 1987. *Glossary of Fertilizers.* Agricole Publishing Academy, 208-Defence Colony Flyover, New Delhi-110024.

Park, K. 1997. *Text Book of Preventive and Social Medicine.* M/S Banarsi Das Bhanot Publishers, 1167-Prem Nagar, Jabalpur-482001, India.

Reddy, S. R. 2000. *Principles of Agronomy*. Kalayani Publishers, Ludhiana, New Delhi, Noida, Hyderabad, Chennai, Kolkata, Cuttack, India.

Ticku, B.L. 1993-94. Monitoring environment through satellite imagery combined with successive ground sampling, 1976. Paper *Dharti*, 1,(3) : 1-6.

WHO. 1997. The World Health Report. Report of the Directorate, Director General, World Health Organisation.

World Bank. 1995. World Development Report. Workers in a integrating world.

INDEX